ALIENS
UFO PHENOMENON

PROF. R V M. CHOKKALINGAM

Copyright © Prof. R V M. Chokkalingam
All Rights Reserved.

ISBN 978-1-63781-054-5

This book has been published with all efforts taken to make the material error-free after the consent of the author. However, the author and the publisher do not assume and hereby disclaim any liability to any party for any loss, damage, or disruption caused by errors or omissions, whether such errors or omissions result from negligence, accident, or any other cause.

While every effort has been made to avoid any mistake or omission, this publication is being sold on the condition and understanding that neither the author nor the publishers or printers would be liable in any manner to any person by reason of any mistake or omission in this publication or for any action taken or omitted to be taken or advice rendered or accepted on the basis of this work. For any defect in printing or binding the publishers will be liable only to replace the defective copy by another copy of this work then available.

This book is dedicated to the ancient cultures or prehistoric civilizations who have given some of the amazing archaeology or spectacular relics or enigmatic constructions or megalithic monuments or marvelous structures or remarkable artifacts around the world that often astounds us for their uniqueness, their beauty, and above all, their highly advanced technology- It leaves us in awe even today, when we take a look at these scientific mysteries of incredible feats of engineering that our ancestors were able to achieve on a scale we cannot comprehend thousands of years ago.

Contents

Preface — vii

About The Book — xiii

1. Alien Dream — 1
2. Alien Life — 8
3. Alien Visit — 17
4. Ancient Alien Theory — 26
5. Alien Evidence — 33
6. Pyramid Technology — 41
7. Alien Search — 50
8. Ufo Phenomenon — 59
9. Ufo Encounter — 69
10. Alien Science — 81
11. Alien Enigma — 93

Bibliograpbhy — 99

Author Bio — 105

Preface

"I am sure the universe is full of intelligent life. It has just been too intelligent to come here"-

Arthur C. Clark

I have passion for UFO and obsession with Aliens from my childhood. I do tend to turn to the night skies more often in search of something distracting. When it comes to Aliens, I think a lot of people are very much interested and personally feel it is hardwired to us. Tales of UFO sightings have been common knowledge over the past 100 years. UFO sightings have been occurring for a rather long time and have indeed, captured the imagination of citizens all over the world. World-weary skyglancers and cell phone-wielding observers across the world have been reporting some mysterious sightings from recent times. Every now and then, we come across a report of a UFO being sighted in some part of the world. UFO phenomenon includes a broad spectrum of human experiences, ranging from simple sightings of unusual lights in the sky to complex extraordinary experiences with unknown beings and objects. There are eyewitness accounts not only from jet pilots, but from radar operators. The Pentagon has released three declassified Navy Videos in July 2020 that have driven speculation about Unexplained Aerial Phenomena.

When a fresh video or picture of UFO sighting surfaces, discussions on social media turn into theories, sparkling reactions from millions of people. UFO and Alien theories pique the interest levels of people around the world. MUFON or Mutual UFO Network is among the biggest and oldest groups with well-trained volunteers collecting UFO sightings- and investigating them in detail for more than 50 years. They have been collecting accounts of UFO sightings largely through a phone hotline and in recent decades on online form. While SETI or Search for Extra-Terrestrial Intelligence continues its research, and astronomers detect more and more exoplanets, a considerable part of the population is not only convinced that Intelligent Alien life exists, but that it has already made contact. These speculations cover a wide range of topics including channeled messages from benevolent space commanders and frightening narratives of being abducted by mysterious Grey Aliens. When we make a mention of the word Aliens, it conjures up images of black-eyed, large-headed, dwarfish beings that have come to be known as Grey Aliens.

Sightings of UFOs of varying sizes, shapes, and other characteristics have ben recorded all around the world, whose identity and happening is a big mystery. Extraterrestrial life comes in all kinds, appearances, sizes, and bodies. The human race has encountered them and continued to do so even today. The Spacecrafts can fly silently, they can do seemingly impossible manoeuvres, and suddenly disappear at an incredible speed. The international conflict is correlated with a cultural fascination with UFO. As humanity ramps up its search for Alien life, we should keep in mind the extraterrestrials might be hunting for us as well. The best we can probably hope for if Aliens are visiting Earth, is that they are studying us. The more Alien Species have in common with us, the more likely they are to visit Earth, if they are able to. Oxygen- breathers would be more likey they are to visit a planet with oxygen atmosphere. Carbon-based life forms would be more likely to be interested in planets suitably for carbon-based life. Electromagnetic creatures that live on Neutron Stars, for example, would be extraordinarily difficult to discover or communicate with.

Fermi Paradox is spawned by the belief the billions of stars must hold life beyond our knowledge. The existence of UFOs and the artifacts left by ancient astronauts confirm a universe filled with intelligent life. In other words, a universe filled with intelligent life explains present UFO visitations. We observe world mysteries such as megalithic monuments, Nazca geoglyphs, or so-called Vimanas of ancient Indians scriptures, Pyramid mysteries, or the vision of Ezekiel- everything ends in question marks. The advanced Et civilization is not an Alien race in the way we normally think of Aliens- they are our ancestors and as humans as we are. Because of their celestial origins, and far advanced powers, earthly humans could only understand them as superhuman divine beings from their primitive perspectives. Traditional theological, spiritual, or psychic explanations of phenomena are replaced by profoundly physical, material explanations. Scientists have greeted the topic with skepticism most often dismissing Ufology as pseudoscience, and believers in UFO and Aliens as irrational or abnormal. The scientific community has underwhelming reaction to new evidence on UFO and Alien life.

But several scientists now acknowledge the discovery of extraterrestrials would be the most important event in the human history. It is no wonder academics, professionals, and scientists publicly shy away from the subject of UFO. No one knows where and when a UFO can potentially appear, hence the difficulty of scientific research with the domain. UFO phenomenon is a scientifically interesting problem, driven in part by observation that seem to defy the laws of physics. At present, there is a rising call for UFO to be studied scientifically, perhaps even using satellites t be on the lookout for possible future events. There are certain aspects of UFO sightings that cannot be explained and we should consider the UFO phenomena worthy of study. UFO community has been developing some compelling theories about the origins and nature of UFO technology. There is also the question of whether living beings are physically piloting UFOs or whether the Unidentified Aerial Phenomena are potentially pilot-less drones. Using multidisciplinary scientific approach to the UFO phenomenon will be what it takes to solve the mystery once for all.

I have always been interested in UFOs with the excitement that there could be Aliens and other living worlds. UFOs have been observed through telescopes and much of the sky has been swept for Alien signals in the form of optical, infrared, and radio waves. For decades, academic researchers have dismissed the study of UFOs as Fringe Science. But as the evidence becomes harder and harder to ignore, some scientists are finally taking steps to make the field legitimate. We are in the midst o a modern ufological renaissance, some renegade scientists are fighting to bring academic rigor to UFO research. A few UFO organizations do remain with their unique contributions to the phenomena in data collection roles and long term scientific study of cases. We need an extensive and well-funded scientific investigation of the UFO phenomenon using state-of-the-art investigative team. Younger scientists are particularly driving the change and are more likely to support the study of UFOs. We are left to wonder at an Alien nonhuman intelligence, or something fundamentally beyond our physical and philosophical understanding. Aliens are really serious science, and this fringe field of science need to go mainstream.

Prof. RVM. Chokkalingam
BE (Mech) CDSE (Lond) Dip (Theo) Dip (Astro)

About The Book

It is ostensibly a book about Aliens, but it gives so much more. It seems like a niche topic and each chapter delivers concise and accessible message. The book invariably observes the UFO arena as a kind of vivarium for a range of psychological, sociological, and anthropological experiences, beliefs, conditions, and behaviors. The beginning of the UFO era dated 1947, and soon after extraterrestrials and their space craft were the most common explanation for mysterious sightings in the sky by the general public. The intriguing book provides illumination, insight, and perspective on the wider UFO phenomenon. It also analyzes, and brings to light anything remotely connected what the world identifies as UFOs, and Extraterrestrial life. The book shares an important and outstanding contribution about the ancient visitors and their influence on our planet Earth. The book explores the fascinating enigmas and mysterious artifacts our ancestors left behind, from incredible objects to amazingly accurate ancient artifacts: from the Great Pyramid of Giza and stone megaliths at Gobekli Tepe, to The Nazca Plains and mysterious structures of Puma Punku. It also provides a captivating adventure around the world and sheds an interesting perspective on the ancient Astronaut Theory. The discovery of independent life beyond Earth would have deep philosophical implications for us and our ideas of morality. Another hypothesis in today's ancient astronaut discourse is the idea that Aliens, because of their superior technology, were taken for divine beings with supernatural powers. We are open to new-age UFO culture and scientists need to participate in research of UFO sightings.

CHAPTER ONE

Alien Dream

" "Facts do not cease to exist because they are ignored"---- Aldous Huxley"

Our Earth- a small planet in a modest solar system, a tumbling pebble in the cosmic stream, lush with living things beyond number: excites the senses and exalts the soul. Life has a way of being obvious, yet it is notoriously difficult to define in absolute terms. And this kind of life has a big question: what else is alive out there? The most tantalizing possibility is that the universe hums with life. The simple truth: extraterrestrial life, by definition, is not conveniently located, and represents an enormous gap in our knowledge of nature. Therefore, it has now become an increasingly exciting area of scientific inquiry. From the Epicurean Greek philosophers more than 2000 years ago to fiction writers of today, people have speculated about the possibility that there might be other worlds which are home to alien life. Despite uncertainties, the science of extraterrestrial life has been infused with optimism in recent years. We do not know how truly alien it is and we do not know if it is built on a foundation of carbon atoms. Nevertheless, we need to begin the long process of putting human existence in its true cosmic context. Life beyond our own planet would teach us things about ourselves we might never otherwise learn.

As we learn more about our own planet and life we feel a stronger urge than ever to put into context. Finding irrefutable signs of any life beyond Earth, whether in our solar system or around nearby stars, would surely be one of the most revolutionary discoveries for human science as well as society. It seems inevitable that there must be other intelligent life out there- it is indeed a very big universe. However, the concept of alien contact is the most popular idea and main interest in science fiction. On a clear and dark Tuesday night in 1953, as a school boy at my birth place Kuppam, a small town, I noticed a falling star in the western sky from outside my house that created nightmare. I was scared about the consequences if something similar happens to our blue planet, which is a lonely speck in the great enveloping cosmic dark. I thought of stars as tiny points of light, but in fact they are huge globes of shining gas. They look so tiny because they are so far away from the Earth. Our Sun is really a star. The Sun belongs to a galaxy called the Milky Way. Aside from our Sun, the dots of light we see in the sky are all light-years from the Earth.

One day in 1962, during my visit to KANGUNDI forest as an adult I observed in the night sky a fast-moving mysterious object with fiery light that suddenly disappeared into clouds. This strange luminous phenomenon caused me to scribble down 1s and 0s in a full page of my note-book subconsciously with a feeling of disconnect. I had sleepless nights for a fortnight pondering over this experience and struggling to understand scribbled strange numbers: probably alien message. If a phenomenon like this seems scientifically inexplicable and if evident for its reality is not available except my narration, that does not mean it has not happened. At the same time no one could tell my claim to have seen UFO like sighting as wrong. But absence of evidence is not evidence of absence. Then my experience extended into the realms of the UFO phenomenon. I became fascinated with the theme of extraterrestrials and the dream of other realms deeply influenced me to learn more about the mysterious nature of alien society: shapes and sizes, patterns of thought, telepathic messages, interstellar travel, and spiritual dimensions. Aliens are hypothetical beings from outer space. UFO is by nature ephemeral.

Reports of weird, wondrous, and worrying objects in the skies date to ancient times. Millions of public all over the world believe we are already being visited by UFOs. UFOs are often linked automatically with extraterrestrials. I think there will always exist the unknowable and the unnamable. I believe there is life in outer space and hope to find the ultimate aliens somewhere in the vast universe. It may not be a surprise to learn that there are 100 billion habitable planets in our galaxy. Complex inorganic molecules from which life develops actually do exist in space. These must surely form or be seeded in other worlds where conditions may later become suitable for germination. I am delightfully speculative that life will be found among the millions of stars. Based on deductive logic most scientists assume that life should be out there in great abundance, but there is mounting concern at our continued failure to find it. Aliens could have evolved surely in different environments and so will have quite different intellectual axioms. Though we do not know for sure, but we want to believe. The probabilities involved in locating aliens call for a substantial effort.

ALIENS

If life exists in any other habitable worlds, it may have entirely different biochemistry from that which we find here on our planet Earth. The mystery of aliens is still preserved. Aliens possibly have been around the galaxy for aeons and encountered many cultures. Of course, we dream about this quest for alien life and still speculate on what forms such beings might take. The UFO phenomenon undoubtedly exists on a major scale and represents data that cannot be ignored. Aliens are now acceptable to our society in such a way that the individual can assimilate. Scientists consider the search for alien intelligence as a purely scientific one, whereas they regard UFOs as a mysterious problem. Some scientists who have devoted time and effort to this subject accept that it is something presently unexplainable. It is very hard to judge imaginative reconstructions against real science because real science has not yet cope with actual aliens of any description. The dream of other realms influences our thinking in a meaningful way. It is possible that a new reality will emerge by embracing the unknown visitors beyond earth or beyond our dimensions of time and space.

It can profoundly affect and forge views of the meaning of our lives in ways that we cannot even predict. The experiences of alien visitation become incorporated into our thinking irrespective of cultural contexts. Its effect would be so sensational that it will let go our fears and help evolve into a higher awareness of ourselves and our place in the universe. It may possibly bring global unification by seeing ourselves as people more of one planet. Any alien contact seems to be an invitation for human development to move from planetary persons to cosmic citizens. It means a neo-religion may emerge based on the aliens in such a way that its gods literally become aliens or otherwise messengers of god and so offer proof that they were here long ago or are coming. This is probable as we mytholize as godly anything that seems to be beyond ourselves. Majority view of science says that aliens are out there and we are on the verge of making contact with a civilization that predates ours by millions or billions of years. Aliens are here, all around us, but we are not yet capable of discovering their existence. Perhaps aliens obscure their presence as they have their own motives.

It is an intelligence that is not here to study us but it is here all along and effectively controls our whole existence. It is doubtless to find that they are like us- because in truth it is we who are like them. Aliens are to be alien in every way and their thought processes are likely to be different from ours. The only remaining question would be whether we could find the ambition to go touch, smell, hear and perhaps even talk to this other intelligent beings. There is an esoteric idea that aliens are psychic invaders within hyper-dimensional space from across the Universe. Within the hidden depths of our mind we grow weary of science and prefer mysticism for answers. God has created man in his own image. God is said to hold domain over other planets and the whole universe. We do not know whether aliens are spiritual, and worship god or embrace that sort of concept. We seek to evolve into a higher awareness of ourselves and our place in the universe. Many philosophers and religions argue that the ultimate purpose of life is to become truly self-aware. Surely all beings in the cosmos are not solely the product of natural processes, but are also the creation of a mystic force, otherwise known as God.

We believe that life exists on many levels and frequencies of energy manifestation. This is why some UFOs are reported as disappearing and then reappearing again. UFO consciousness- a state of mind in which the Earth itself brings to us and introduces unidentified into our midst. UFOs seem to be the product of an inner reality representing outer truth. The human mind does suck into any viewpoint it desperately wants to be true. When we enter into research with a belief system already intact, it can often taint everything we see. Were we been visited? Were our species engineered? Were our intellect been artificially stimulated by alien visitors from another world in the distant past? Were they responsible for making visible signs everything from earthquakes to crop circles to UFOs? These questions are really worthy of deeper exploration by researchers and scholars as they remain unsolved mysteries. We live in a universe that contains inhabited worlds beyond imagination. Also alien race's way of thinking seems to be utterly alien to human modes of thought. Our civilization has to grow up in wisdom only by embracing what lies beyond and by exploring our vast universe.

The UFOs exist, the proof is undeniable. The UFOs have become a living myth. The very shape of UFO, usually a disc or globe reminds us the Mandala or Magic Circle, which has been the universal symbol of order and perfection throughout the world since time immemorial. It is also the symbol of God and the higher self. The fascination with the UFO, the experts regard as a continuation of the notion that the spirit and the giver of the spirit, and the power and the glory are out there somehow and not here within us. Space people are more likely a product of the inner space than outer space where there might be real aliens. We dream about the quest for alien life, and, still speculate on what forms such beings might take. All the mathematics say there must be someone out there and physical contact with alien race is a genuine possibility in the future. Everything on Earth evolved from a single ancestor. But when did life begin? Where did it come from? How did it form? These are undoubtedly the biggest questions in biology, harder to answer, and the most controversial. We are still embarrassingly short of data on how precisely life got going on Earth.

We ourselves are part of nature and therefore, part of the mystery that we are trying to solve. The mystery of why we are here is perhaps greater than ever. Many suggest that the aliens seeded the human race and we are the descendants of the aliens. The aliens may not speak that part of our consciousness that we dream most important- our spirit. However, the aliens seem as intangible as ever. Contact with an alien civilization would be an epochal and culturally challenging event. Alien contact could give humanity the cultural boost to re-examine our basic tenets. Aliens live deep into our collective consciousness mirroring the good and evil we are capable of. The concept of extraterrestrials intelligence has become part of modern culture. And most UFOs often turn out to have banal explanations. The idea that aliens are extraterrestrial sentient inhabitants, the cosmos is surely more popular today. Nobody knows where UFOs come from and somebody somewhere knows the real truth. Space is awfully big, so it might seem just too hard to detect some distant alien civilization. Sure, it is hard to zoom across the vacuum of space.

CHAPTER TWO

Alien Life

"So where is everybody?".... Enrico Fermi

Our universe is about 14 billion years old, while Earth is about 4.5 billion years old. Looking through a telescope into space we can see vast numbers of stars and galaxies. Our galaxy alone contains several billion stars. Stars are nothing other than suns. In reality many stars are considerably larger than our Sun. if we can see the Milky Way from the outside, It will look like a gigantic wheel of stars. There is a deep-rooted, perhaps intrinsic belief in us that there is something out there in the interminable vastness of outer space. The real world, indeed, holds wonders far more mystifying and incredible than anything our mind could ever have conjured up. As we look out into the seemingly infinite heavens, our thoughts shift from the daily trials of life to questions of who we are? How we got here? And why we exist? Religion, science, and philosophy essay to answer these questions, and often the answers seem incomplete and unsatisfactory. We still do not know how life on Earth got started and it is still possible that we never will. The existence of life, with all of its biochemical complexity remained beyond scientific understanding. All life must have gotten to start somewhere.

The question of life beyond Earth makes sense only if we have reason to think that we are made of star dust- we are star stuff. It implies that almost every atom in our body and in our planet Earth was made inside a star. As we study the universe as a whole we realize that the microcosm and the macrocosm are increasingly the same subject. Each man is a microcosm of the universe. Our body is made up of all the elements of the world. Man is the product of the attributes of Heaven and Earth by the interaction of the dual forces of nature and the finest subtle matter of the five elements. The universe made us by donating itself. The concept of a universe designed for life must surely encompass the universality of life throughout the vast universe. It is highly improbable to think that we are alone in that enormous immensity. Stephen Hawking says: with 100 billion galaxies in the universe it seems perfectly rational that aliens exist. Earth is so unique among the known worlds. The fact of our own existence makes it philosophically reasonable to wonder if life exists beyond Earth.

Earth makes it reasonable to think that life would emerge just as quickly on other worlds with similar conditions. Physicists today recognize that the fundamental laws are finely tuned to enable the evolution of a universe in which organic life can develop. The same laws of nature operate throughout the universe as here on Earth. As we learn more about solar systems, we begin to realize that other planetary systems probably are not uncommon. It seems plausible to imagine that other stars could have their own planets perhaps with life. It is eminently reasonable to think that life could be quite common on worlds that number beyond imagination. A world can be habitable if it offers the potential to support life. Scientists have been engaged in the search for Earth- like planets to investigate potentially life supporting environment. The environmental requirements for life are a source of molecules, a source of energy and a source of liquid water. The life creates order out of chaos, but needs molecules to serve as building blocks. Molecules build its own cellular structures and reproduce itself.

Living organisms can take fairly random assortments of molecules and arrange them into orderly patterns that make cell structures and govern the metabolism of life. The life needs energy to maintain biological order and to fuel chemical reactions that occur in life. Living organisms use energy to create the order upon which their survival depends. The life would decay into disorder without energy input. The life requires liquid water to move molecules around both within a living organism as well as into and out of the organism. Life evolves through time by making it possible for species to adapt to changes in the environment around them. Creating order out of chaos is implicit in the ability to reproduce, which in turn is implicit in the fact that life is able to evolve through time. Scientists have found organic molecules in meteorites and in gas clouds between the stars. They try to tell that the microbial life might prove to be common throughout the universe. All known life on Earth uses DNA as its hereditary material. We cannot tell about the biochemistry of Aliens nor about their cells use DNA as a genetic material.

According to academics, our ability to humanize the aliens is nothing more than anthropomorphism of perceptible creatures. The chemical content of the universe seems to consist almost entirely of just two elements: hydrogen and helium. These two lightest and simplest elements make up at last 98 percent of the matter found in all stars and all gas clouds in space. All the rest of elements form carbon and oxygen that make up a large portion of our bodies to the gold and silver, which are no more than 2 percent of the overall chemical content of the universe. The life on Earth is carbon-based: all the important molecules of life including proteins, fats, carbohydrates and DNA are essentially long chains of carbon atoms attached to various other atoms such as hydrogen, oxygen and nitrogen. These atoms are star stuff existing throughout the universe. Life on Earth is based on the uniquely versatile chemistry of carbon. But none of this diverse and complex biochemistry could function without a solvent with unique properties: water. This seems a way to look for worlds that have a carbon source, available energy and a liquid water medium.

Scientists are now trying to understand life's origins into questions such as when, where, and how did life arise on Earth? Studies of asteroids and comets reveal that they contain lots of organic molecules, including complex molecules such as amino acids, which would therefore have come to Earth with every impact of the heavy bombardment. Therefore, Earth should have received abundant organic molecules from space. We do not know that we are nothing more than random reactions of atoms in a universe without purpose or miraculous creation of God. With the present understanding of life and evolution, we have no scientific evidence against the role of God. Proponents of Intelligent design claim that life is so intricate and complex that it could not have arisen naturally, and they therefore claim that life must have been deliberately designed by an Intelligent Designer- life is the work of a supernatural designer which is beyond our scientific comprehension. Something remains hidden which cannot be discovered solely by scientific investigation. It is linked with that great mystery concealed at the very heart of the universe for which science has as yet no answers, and it is the secret of life itself.

Our science has deposed the gods, but we retain the psychological need for them to exist and to intervene in our lives. The more we explain the universe in scientific terms, the more pointless it becomes. But that is because science has left a key element out of the equation- the human soul. Our soul is the formless life force, the consciousness that brings alive our human form. Science cannot give us the meaning of life. In essence, the universe appears to be as it is, because it must be that way, its evolution was written in its beginnings- in its cosmic DNA. It has long been known that the universe consists of, quite literally, countless billions of stars. All the stars share certain characteristics. They are considered to be gigantic nuclear furnaces which burn their fuel to create heat and light. Based on various factors such as size, power, age, they range from very hot giants to almost cold dwarfs. Our Sun happens to be a middle of the road star, which means it is of average size and intensity. Stars are not living organisms, but they nonetheless go through life cycles. All stars are born from the gravitational collapse of large clouds of interstellar dust and gas.

A star is born when its core becomes hot enough to start the fusion reactions that power it throughout its life. The star shines until it ultimately runs out of fuel for fusion, at which point it dies, scattering some of its remains back into space, where they can be recycled into later generations of stars, while leaving the rest behind as a stellar corpse that may either be a white dwarf, a neutron star or a black hole. Though all aspects of stellar lives are fascinating, the black holes still test the limits of our understanding of physics.

Not all stars could even conceivably host intelligent life. Those that are too hot or too cold or too unstable would almost certainly have never had the right conditions for a sufficiently long period to allow life to develop. Stars similar to our Sun in mass clearly qualify as potential suns, since they would have lifetimes of many billions of years and habitable zones as wide as that of the Sun, offering ample room for the existence of Earth-like planets. But we do know that the basic building blocks from which the stuff of life is formed are found all over the universe. We also know that primitive organisms can exist even in space itself. Yet we do not know how many suitable stars have planets.

It would be totally invisible to the biggest telescopes in existence, as the distances are so great and the planets so relatively small that they are impossible to see. Also light from the star itself is so enormous that it would completely swallow up any light from planets circling around it. With billions of planets out there, the odds are that some of them will be Earth-like in form around sun-like stars. Major new telescopes may expand the search for Earth-like planets, and investigate potentially life-supporting environments. Let us look at our cosmic address: we live on a planet, Earth- that is the third planet out from the star that we call the Sun in our Solar System. Our Sun, in turn, is one of a vast collection of stars that make up what we call the Milky Way Galaxy. All the stars visible with the naked eye are part of the Milky Way. Our spiral galaxy belongs to Local Group of Galaxies. The galaxies are separated from each other by enormous distances. Most galaxies reside in groups, which are called Clusters. When Clusters are grouped together they become Super-clusters of galaxies. Together, all the Super-clusters and all the spaces between them make up what we call our universe.

Our recent progress has made the possibility of life elsewhere probable. With our scientific quest, we may seek a deeper fundamental reality by embracing mysticism. What today's mystical nonsense may often become tomorrow's accepted science. The existence of pattern, purpose, and design throughout the universe is the consensus of the world's spiritual traditions. Earth has made it habitable because it is of the right size and within the right range of distances from the Sun. Current scientific understanding makes it reasonable to think that life could be quite common on worlds that number beyond imagination. Millions of people right across the world believes in the theory that they have been visited by extraterrestrials. It is an investigation into a theory some believe cannot be true, but many agree cannot be ignored, yet others feel hard to dismiss. The aliens are, by definition, separate from us: yet they are also within us, part of our inquisitive mind If there is life out there. When an alien consciousness resides within you and without you, fantasy and reality become meaningless terms. Some claim that alien intruders are within our minds- psychic invaders from across the universe.

There is a growing social acceptance that we are not alone and may be inwardly convinced of that already. The human race is so fascinated with the theme of alien existence. What is an alien dream for one person is reality for another. Fear is one of key emotions in the alien equation. The aliens seem as intangible as ever. Space scientists tend to believe that alien life can take any form, from primitive single-celled entities to strains of bacteria that can survive under the most hostile conditions, to actual sentient beings. To them the word 'aliens' seems to describe the creatures living somewhere else in the universe, separated from us by unimaginable distances. There remains the centuries-old idea that life exists on other planets, and that humans and alien beings have crossed paths before. Also there exists a controversial theory often maintains that humans are descendants of beings who landed on Earth thousands of years ago. They created modern mankind by mixing their genetic make-up with that of sub-humans. A form of genetic manipulation seemed possible blending humanity with extraterrestrial genetics. Some scientists wonder whether if life itself was seeded on Earth by an alien civilization.

Many researchers believe that ancient aliens might have seeded Earth, left their offspring to evolve on Earth, and eventually led to the race we now call human beings. There was also the possibility that alien beings might have produced hybrid beings with earthlings. An alien is a being whose life does not originate from Earth. A new research in astrobiology supports the view that human life started from outside Earth. The first 'seeds of life' were deposited on our planet Earth from space several millions of years ago. Microbes from outer space arrived on Earth from comets, which then multiplied and seeded to form human life. Humans and all life on Earth, came from aliens brought to the Earth by comets hitting the planet. It is likely that we all on Earth are aliens: we share a cosmic ancestry. However, there is no way to prove or debunk these theories at present. The knowledge of aliens will help us understand the role we play in the universe as well as the meaning of our existence in the entire cosmos. The reality of man's nature and the reason for his existence still remains a mystery despite the enormous advances made in modern science.

The outer edge of alien contact embraces spiritual side of the phenomenon-the links between human potentiality and a universe of aliens, spirits, demons and sprites. But it is nevertheless a motivating force for those who pursue alternative realities of the human psyche. There is an emerging consensus that the detection of alien civilization is a discovery with profound implications for all humankind. We may seek a new reality that embraces the possibility that we have visitors beyond Earth, or beyond our dimensions of time and space. Perhaps we all hope to find the ultimate aliens somewhere. Logical analysis of the facts demonstrates not only the existence of advanced humanities and life on other planets, but their presence on our planet. Any alien would be totally alien, probably to the point that we would have no way of deciphering their motives. The aliens are thought to be beings superior either in physical strength, technology, or spiritual attainment. Scientists suspect that there may be dozens of alien civilizations lurking not too far from Earth and advanced enough to communicate with us. We expect that civilizations older than ours will be scientifically savvy and hence technologically advanced.

CHAPTER THREE

Alien Visit

"Extraordinary claims require extraordinary evidence"—Carl Sagan

Mythology is the study and interpretation of often sacred tales or fables of a culture known as myths or the collection of such stories. Mythological thinking has been part of the human experience ever since the beginning. Mythology is a stimulant that awakens the archetypal forms of a culture in the form of stories, rituals and symbols. Mythology exalts the divine revelation and infuses it with sacred and mystical significance. The gods and goddesses of ancients were creations of collective unconscious. Mythical images point to dimensions of our own inner life, of the inner life, releasing their energy and wisdom into our lives. The stories of mythology may not be literally true, but they are metaphorically and existentially true. Mythology is the code that contains the record of the journey of those earlier generations. Many ancient myths incorporated ideas as they often imagined mortals joining the gods among the stars. Mythology contains a mystical power where the sacred can enter our world. The externalized symbols of archetypal images and universal energies reside in the deeper layers of the human psyche. Mythology, therefore, contributes to the spiritual and existential wisdom.

ALIENS

Many Sanskrit epics of ancient India written in and around 3000 BC, contain references to mythical flying machines called Vimanas. These oldest writings of Mahabharatha and Ramayana speak extensively about flying machines that flew not only to the skies, but to the stars as well. They describe at length the chariots powered by winged lighting- it was a ship that soared in to the air, flying to both the solar and stellar regions. Vedic writings of India speak of Gods or Devas from other star systems. The Vedic Gods are not simply poetical personifications of abstract ideas or of psychological and physical functions of Nature, but they are living realities. They explain how human soul is nonphysical and how our sentient energy goes through a succession of physical bodies, as only the physical body dies. Sanskrit texts are filled with references to gods who fought battles in the sky using vimanas equipped with weapons as deadly as modern ones or even fantastic science fiction weapons. Brahmastra was embedded with the mystical force of Brahma, which releases innumerable missiles and great fires with a destructive potential to distinguish all creations in mythical battles.

Myths tell the stories of ancestors and the origin of humans and the world, the gods, supernatural beings and heroes with super-human, usually god-given, powers. The writings in Mahabharatha describe weapons reminiscent of guided rockets, beam weapons, and nuclear devices used by the gods. VEL was the spear-shaped weapon of the Hindu war God Murugan. The Vel in Tamil was the divine spear that was a powerful and infallible demon-slaying weapon. The sacred Vel of Lord Murugan symbolizes penetrating spiritual knowledge, wisdom, and the removal of ignorance. Vel was the symbol of valour and the triumph of good over evil. Mythology is the repository of man's most ancient science. Myths, legends, fables, fairy tales, epics, all tend to shade into one another and little difficult to distinguish. We must recognize that myths convey cosmic and anthropological facts, just as fables convey truths about the social behavior of men and women, and fairy tales reveal one aspect of the human subconscious or the psychic nature of every man. If we dig deep enough, say scientists, we can find some truth to legends and creation stories.

We find that in Egypt, especially, there were many gods with heads of birds or animals. Thus, Hermes, the God of Wisdom was also associated with Hermanubis and Anubis. Thoth was the god of wisdom and of authority over all other gods. He was the recorder and the judge. Anubis, the Egyptian god of generation was represented with the head of an animal, a dog or a jackal. When Egyptians represented the Moon as cat, they simply saw that the cat sees in the night, and in the dark her eyes become luminous. In the heavens, the Moon is the seer by night, reflecting the sunlight, and so the cat was adopted as a representative of the Moon. Legends spoke of wars between gods and titans, men and beasts, of powerful deities who ruled the universe with their might and heroes who rose against monsters and the supernatural. Stories revealed about the names of the powerful weapons these mighty beings used to rise to the tops of pantheons and uprooted the most powerful of enemies. In Greek mythology many weapons appeared in the stories. They varied from the staff of Hermes, the caduceus to the head of Medusa.

The Trident was the signature weapon and symbol of Poseidon. It had the power to control the seas, create earthquakes, and cause storms. The Aegis was the name of the shield Zeus and Athena used. It was also a breastplate.

It radiated fear, turned others to stone. Zeus was the lord of the sky, and the rain. His weapon was a thunderbolt, which he used to hurl at those who displeased him. The caduceus was the weapon of Hermes, which gave the golden touch, transforming items and living beings into gold. The divine titan Prometheus stood for spiritual creators and also represented fire by friction. Mythology with a fuller consciousness of its origins and functions is a tool for exploring the deeper dimensions of the soul. Mythology has always been about the inner life and it projects into the world our deepest archetypal patterns. Ancient Egyptian legends tell of Tep Zepl or the First Time: an age when sky gods came down to Earth and raised the land from mud and water. Like other creation myths, Egypt's is complex and offers several versions of how the world unfolded. The ancient Egyptians believed that the basic principles of life, nature and society were determined by the gods.

Sumerian writings speak of how man's early history was greatly influenced by visitors from the twelfth planet in our solar system. The Sumerians believed they were created by the Annunaki, who came from the stars to planet Earth to mine for gold. The Annunaki genetically engineered and created the Sumerians as they needed workers to mine for gold. The Sumerian Sar-Ur weapon of Ninurta was better known as 'Flood- Storm' weapon. It was described as the ultimate weapon, which could fly and there was no resistance to this storm, a falcon against the foreign lands whose wing bears the deluge of battle. In his heart he beamed at his lion-headed weapon, as it flew up like a bird, trampling the mountains for him. It raised itself on its wings to take away prisoner the disobedient, it spun around the horizon of heaven to find out what was happening. It also gave off great radiance, and brought forth light like the day. It could also speak, and was capable to bring in fire, deluge and poison. It was the ultimate weapon that could obliterate mountains. Mythology reflects images in the depth of our own souls that speak to the universal concerns of the human being.

The Mayans believed their predecessors came from the Pleiades. The Popol Vuh states that several gods including Hunahpu, Xbalanque and the great god-king Ouetzalcoatl returned to the stars after their earth life ended. Mythology was founded on natural facts. To understand a myth is to understand its purpose, its significance, and its practical consequences. Mythology opens us to the unfathomable depths of our own spiritual nature and inner life of the soul. Most of the ancient civilizations had beliefs in gods that interacted with or created them. These gods in most cases were from the stars or sky. These gods with their supernatural powers taught them, guided them, and ruled over them. Many indigenous tribes and races also believed that life was brought here by 'serpent gods from the skies' or other godly creatures. Some religions thought god was an alien, so perhaps an alien visit was seen as fulfillment of prophecy. The Aztecs of Mesoamerica believed that their gods needed human hearts and blood in order to remain strong. Their sun-god, Huitzilopochtli, was just one of many gods and goddesses who demanded bleeding hearts and human sacrifices on a regular basis.

Prisoners of war became the main source of these sacrifices, which instilled fear in the hearts of their enemies. Mythological symbols interpreted literally support civilization. Thor was the god of thunder and of the sky in Norse and early Germanic mythology. His weapon was mighty hammer or crusher which when thrown, crushed the giant' head and returned magically to his hand like a boomerang. Its other power was the gift of restoring life to the dead. Inca mythology included many stories and legends in which Viracocha was the supreme god and creator of all things. Apu was a god or spirit of mountains who received sacrifices to bring out certain aspects of his being. Throughout history fear of the unknown has been the basis of belief in god or the gods. Despite advances in science, for most people the unknown still remains frightening; too many things are beyond human probing and speculation. Our scriptures declare that heaven and earth are full of god's glory. According to legends, King Solomon of Israel had a flying machine that enabled him to travel great distances and make maps of the world. There appeared evidence that the ancient Egyptians had been visited by beings from star Sirius.

According to mythology, the Dogon people of Africa had a whole range of astronomical knowledge brought to the earth by the 'Nommos'- an amphibious race of aliens from the Sirius star system. The Dogon described the landing of the Nommos 'ark': The ark landed and displaced a pile of dust raised by the whirlwind it caused. The violence of the impact roughened the ground. He was like a flame that went out when he touched the earth. The Nommos: The whole body of the animal was like that of a fish; and had under a fish"s head another head, and also feet below, similar to those of a man, subjoined to the fish"s tail. His voice too, and language, was articulate and human; for he was an amphibious to plunge again into the sea. All over the Middle East, stone reliefs and drawings abounded of beings adorned with fish tails. One of them, a Babylonian semi-demon called Oannes, was said to have founded the first civilizations on earth. In the book of Ezekiel, part of the Hebrew Bible, the prophet Ezekiel had a vision of a flying vessel accompanied by fire, smoke, and loud noise. Ezekiel described his encounter with beings who seem to descend from a 'wheel of fire' turning in the sky.

The 'living creatures' had the appearance of burning coals of fire, and the appearance of torches, who ran and returned as the appearance of a flash of lightning. The living creatures confronted Ezekiel with prophecies of the future. Ezekiel then saw a vision of god seated on a throne, who instructed him on a program of reform. This enigmatic piece of writing by Ezekiel with the utter complexity of vision is definitely something totally outside the experience of early humans. This possibly suggests an eye-witness account of nuts-and bolts flying machines of extraterrestrial origin in the form of some sort of helicopter. Ezekiel's repetitive and detailed description of the 'wheels' had the effect on the design of modern space vehicle. Many episodes in Old Testament can be seen in terms of a period of history where alien visitors took it upon themselves to meddle with the genetic and cultural evolution of early humans. It is hard enough to comprehend that the aliens operated on many frequencies; they were able to manipulate matter, and occasionally operated on our physical plane. On each of these levels there were nuts-and-bolts apparatus, including spacecraft to travel and weaponry to fight off the evil.

They constructed physical objects on whatever plane they chose through advanced scientific methods, including thought control. Perhaps they beamed thought power across the universe and established contact through a method unknown to our scientists. In 1968, the Swiss author Erich Von Daniken published Chariots of the Gods? But the book, along with its successors, became an international best seller, and Erich Von Daniken, a household name. In it he did profess his hypothesis that thousands of years ago, space travelers from other planets visited Earth, where they taught humans about technology and influenced ancient religion. Erich Von Daniken is regarded by many as the father of ancient alien theory: also known as the ancient astronaut theory. Ironically, despite the book's controversial claims, it does not state that God was an astronaut. Ancient alien theory grew out of the centuries-old idea that life exists on other planets, and that humans and extraterrestrials have crossed paths before. Many ancient alien theorists point to the ancient religious texts as evidence, in which humans witnessed and interacted with gods or other heavenly beings who descended from the sky in vehicles or spaceships.

Interestingly the religious texts spoke of visitations from the Heavens by powerful beings who performed extraordinary miracles and then left. Some ancient alien theorists claim that these aliens form the basis for the mythological gods of ancient culture. They propose that deities from most religions are actually extraterrestrials and their use of technologies were taken as evidence of their divine status. Based on ancient legends or stories, they suggest that the development of various weapons through history might have been guided by aliens. Thus mythology with descriptions of gods or demons coming from the sky has been used as evidence for ancient extraterrestrial visitation by proponents of ancient alien theories. The supernatural powers were merely the interpretation of their advanced technologies. If we believe in alien visitation in the past, ancient texts were surely influenced by alien encounters. Some people today regard these beliefs, held by those ancient civilizations, as just simply myths or legends. Others feel it purely subjective that aliens were here and affected the history of the Earth.

According to ancient alien theorists, most of whom have researched the topic for decades, based on a few clues scattered throughout the world, alien beings could have very likely visited the Earth long ago. To the question if aliens visited Earth in the past, could they make appearance in the future?- the answer from ancient alien theorists is a resounding yes. Some skeptic camp took up the challenge of ancient alien theory and commented wildly on theories suggesting that astronauts wandered the Earth freely in ancient times as hugely speculative and expounding. Some members of scientific community criticized many ideas presented in the ancient alien theory as pseudoscience and pseudohistory. Any objective reading of these theories fills one with feeling of unease as it suffers for a lack of scientific credibility. The idea is not taken seriously by most academics, and has received no credible attention in peer reviewed studies. But proponents of this idea present evidence in favour of their beliefs, it is often distorted or fabricated. True believers and skeptics rarely go over to the other side. The aliens that we so dearly seek to contact are proving both frustrating and elusive.

CHAPTER FOUR

Ancient Alien Theory

"Of course it is possible that UFOs do contain aliens as many believe"—Stephen Hawking

Ancient aliens or ancient astronauts are purported intelligent extraterrestrial beings said to have visited our Earth in antiquity or prehistory. Ancient alien theory or ancient astronaut theory grew out of the idea that life exists on other planets, and humans and extraterrestrials have crossed paths before. The ancient alien theory, also known as Paleocontact hypothesis is the thought that thousands of years ago extraterrestrials visited our beautiful planet Earth and introduced new form of knowledge and science to population to help with our advancement as a society. Proponents suggest that aliens made contact with humans and influenced the development of human cultures, technologies, and religions. A common variant of the idea is that deities are actually extraterrestrials, and their advanced technologies were wrongly understood by primitive men as evidence of their divine status. Historic accounts from Egyptians, Sumerians, Mayans, Aztecs and many other ancient civilizations mention strange phenomena that came from the sky like our modern day UFOs. Alien encounters had been documented in various historical texts, and medieval art depicting disc-shaped objects floating in the heavens, as evidence.

Most alien theorists point to ancient religious texts in which humans witnessed and interacted with gods or other heavenly beings who descended from the sky-sometimes in vehicles resembling spaceships- and possessed spectacular powers. They also point to the physical specimens such as artwork depicting alien-like figures and ancient architectural monuments. So much of these theories are dependent upon ancient writings, paintings, and structures. The Great Pyramids, the Nazca lines, the writings of Mesopotamia and countless other structures and reliefs all provide valuable insights and prove previous alien visitations and interventions. The ancient alien theorists theorize that extraterrestrials with superior knowledge landed on Earth thousands of years ago, sharing their expertise with early civilizations and forever changing the course of human history. Seemingly many supernatural events reported in the ancient texts, the megalithic constructions, the magnificent architecture, the enigmatic lines and numerous artworks over the Earth resulted from an advanced technology, which were incomprehensible and indescribable by the ancient human observers.

They did not possess necessary tools to create some of these marvelous structures and monuments. So, they wondered how did they do it? Where did they get this knowledge? And why there was no record for it? The perfection and the way some of these structures were built show out they had definite help from ancient alien visitors. The ancient alien theorists argued that the archaeological artifacts were anachronistic or beyond the presumed technical capabilities of the historical cultures with which they were associated: sometimes referred to as out-of-place-artifacts. The ancient civilizations built huge monuments that stood the test of time, which even with today's technology and machinery would be hard to duplicate. However, the existence of structures and artifacts have been found representing higher technological knowledge than is ever presumed to have existed at the times they were made; these artifacts were produced either by extraterrestrial visitors or by humans who learned the necessary knowledge from them. They built the megalithic structures and suddenly abandoned some of them. They even at times, obliterated man and started all over again.

Ancient alien theories have been popularized in the later half of twentieth century, pioneered by Erich Von Daniken, a Swiss autodidact is now considered as the father of ancient alien theory. He was barely nineteen years old when his curiosity first drove him to Egypt to crack down cuneiform inscriptions. He intrigued and fascinated millions of people world over by his bold theories, and daring speculations, and fresh interpretations about mysterious visitors from outer space in ancient and prehistoric times. The top ancient astronaut experts in the world include: Erich Von Daniken, Giorgio A. Tsoukalos, Zecharia Sitchin, Robert KG Temple, David Icke, Peter Kolosimo, David H. Childress, Michael A.Cremo, Jason Martell, Dr. Luis E. Navia, Barry Downing and many others. Though the ancient alien theories suffer for its lack of scientific rigour, they are indeed thought provoking. But the idea that ancient astronauts actually existed as well as visited, has not been seriously taken by most academics, and has received little or no credible attention. Of course, ancient astronauts have been widely used as a plot device in science fiction.

Science ought to show more interest in the less conventional evidence, but the likelihood of success is feeble through standard means. Another associated idea was that much of human knowledge, religion, and culture came from extraterrestrial visitors in ancient times. Ancient astronaut theorists believed that by sharing their views with the world, they could help prepare future generation for the inevitable encounters that awaited them. Ancient legends and stories held the key to unlocking the ancient alien mystery. Powerful gods and monsters were said to share similarities even though these legends were found in different cultures separated by vast distances; it suggested that these legends might be eye-witness accounts of alien visitations. Ancient legends that spoke of fire and secrets of metal works as gifts from gods; stories of divine weapons such as Excalibur as possibly powered by alien source; the mystery of Greek fire as gift from angels for the Byzantines; the development of the complicated chemistry of gun powder by ancient Chinese; the legends of creatures such as the Chimera and Hydra; and humanoids such as the Minotaur and Medusa through alien animal/human hybrid experimentation.

The Anunnaki were the gods as described by the Sumerians, and they had 'Android Beings' helping them. Today's modern UFOs and alien contacts being reported have a strong similarity to the ancient description of the 'Anunnaki' Android beings. Proponents often cited ancient mythologies to support their viewpoints based on the idea that ancient creation, myths of gods who descended from the heavens to Earth to create or instruct humanity were actually representations of alien visitors, whose superior technology accounted for their perception as gods. Erich Von Daniken, a leading ancient astronaut theorist, claimed that the deities of ancient cultures were an attempt to explaining alien visitation. Genetic engineering, cloning, and hybridization technologies could have been used by ancient aliens in the past to manipulate mankind and environment. The creation myths of each ancient civilization discussed alien gods who descended from the sky for any number of reasons, some of whom allegedly mated with woman to create bloodlines or created humans through biogenetic experiments. Science may assess the data as nonsense, whereas in effect the facts have actually been not assessed.

Proponents of ancient astronaut hypotheses often maintained that humans were either descendants or creations of extraterrestrial beings on Earth thousands of years ago. There was awe at what might lie beyond the blue sky and different forces from out there occasionally seeped into human activities. In Hindu mythology the gods and their avatars travelled from place to place in flying chariots. There were many mentions of these flying machines in the Ramayana. Many ancient alien theorists pointed to physical specimens such as artwork depicting alien-like figures, ancient architectural marvels like the stonehenge, the pyramids of Egypt, and Mayan pyramids, and ancient sites like the Nazca lines as evidence to the existence of ancient aliens. The great monolithic gate of the Sun at Tiahuanaco in South America: a gigantic sculpture carved out of a single block of more than ten tons with forty-eight square gures in three rows flanked a being, represented a flying god. They theorized that extraterrestrials from elsewhere in the universe were from a much more technologically advanced and older culture than the humans.

They had put forth their hypothesis that the technologies and religions of many ancient civilizations were given to them by space travelers who were welcomed as gods. This also helped explain the advancements in education and technology to build these amazing structures built by ancient civilizations. More and more evidence showed that ancient civilizations had an amazing understanding of astronomy, science, and mathematics that they were supposed to be one step out of the stone-age. One of many rock paintings which date back to around 6000BC represents an alien in spacesuit. A few claim that the ancient astronauts used the planet Earth a place for their science projects. In Delhi there stands an ancient iron pillar exposed to weather for more than 4000 years without a trace of rust. In this visible time-defying monument, we find an unknown alloy of antiquity cast into a column. In Inyo county of California, a geometrical figure in a cave drawing is recognizable as a normal slide rule in a double frame. According to ancient alien researchers, the proof of alien visitations in antiquity is found prevalently in patterns of art, technology, and engineering.

To them the Moai statues on Easter Island; the Pyramids of Egypt; and the Stonehenge of England can all be easily explained through the ancient astronaut theory. Frescos throughout Europe reveal the appearance of spaceships in the skies including certain painting in 1350. All the gods who are depicted in cave drawings in Sweden and Norway have uniform indefinable animal heads. Ancient alien theorists assert that the primitive people had seen them on the unfamiliar gods. There have been numerous submerged cities found, thousands of years old. Humans could not have designed and constructed them. One of the most important historical sites relating to ancient Egypt is the complex of Abydos, an ancient city. Within the temple of Seti-1, on one of the lintels of the outer Hypostyle Hall there remains a series of carvings that look very much like helicopter and futuristic spacecraft. Ancient astronaut theorists predict that the Egyptians either did whiz around in strange futuristic craft or they did just witness something they could not explain, and so carved it on stone as a record. They interpret ancient artworks throughout the world as depictions of astronauts, their vehicles, and complex technology.

The Mayan- Popol Vuh states men came from the stars, knowing everything and they examined the four corners of the sky and the Earth's round surface. Ancient alien theorists wonder when and how Mayans got their advanced knowledge. They argue that the technological advances in antiquity are indeed the result of alien visitation. They cite abnormal historical surges in technology and unaccounted for invention as proof of their claims. The ancient astronaut theory suggests that the motive for alien visitation is to influence the future of mankind. Though many dismiss these claims as weak, they do provide ample motive for a supposed alien visitation. No matter how we look at it, alien visitors to Earth must be thousands or ten thousands of years beyond us in technology. Even if they are genetically engineered organisms, perhaps they have discovered laws of which we are unaware. It is stated that the most sensational thing that could happen, in the history of man, would be contact with an extraterrestrial civilization. It is extremely ignorance on the part of some who dismiss all ancient alien theories as hoaxes and science fiction babble as we live in a universe where anything is possible.

CHAPTER FIVE

Alien Evidence

"Yes- most likely they are out there, perhaps even visited, perhaps on our Moon"—Michio Kaku

Archaeology studies human activity in the past, primarily through the recovery and analysis of the material culture and environmental data that they have left behind. Excavation is mainly the exposure; processing and recording of archaeological remains. An artifact is an object recovered by some archaeological endeavour, which may have a cultural interest. The systematic study of past human history and prehistory is achieved through the excavation of sites and the analysis of artifacts. We learn more about past societies, their culture and the development of human race by the recovery and examination of remaining material evidence such as monuments, structures, stone giants, rock drawings, cave paintings, sculptures, frescos, statues, seals, figurines, reliefs, relics, tools, pottery etc. Ancient sites provide our imagination free rein to decipher many riddles of our past life and culture. Radiocarbon dating has been used for determining the age of a prehistoric object by measuring its radiocarbon content. All living things contain radiocarbon (carbon 14), an isotope that occurs in a small percentage of atmospheric carbon dioxide as a result of cosmic ray bombardment.

After an animal or plant dies, it no longer absorbs radiocarbon and the radiocarbon present begins to decay or break down by releasing particles at an exact and uniform rate. This made it useful for measuring prehistory and events occurring within the past 35,000 to 50,000 years. For ancient alien theorists there exists a more dramatic situation, one in which extraterrestrial life is not microbial and slimy but rather intelligent, technological, and lurking in our midst. To look at ancient structures built thousands of years ago; how could such architectural wonders that remain standing today? This is a real question that archaeologists and historians have sought to answer by interpreting relics and by analyzing ancient texts. Ancient astronaut theorists believe that the real explanation for such human achievement is through a little "out of the world" help from some distant alien friends. They do suggest that such great things would be impossible to build without today's technological know-how. With the first human societies evolved from primitive hunter-gatherers, ancient artisans could not have the capability to build such amazing structures. So, ancient alien proponents believe them as proof of early alien intervention on Earth.

PUMA PUNKU STONE RUINS: Puma Punku is a large temple complex or monument group that is part of Tiwanaku site of Bolivia in South America. It is an amazing technologically advanced site of megalithic structures believed to be built about 10,000 BC. It is a field of stone ruins scattered with enormous, finely carved stone blocks. The megalithic stones found here are among the largest on the planet. It contains the largest stone slab of 7.81 meters long, 5.17 meters wide, and average 1.07 meters thick weighing about 131 tons. Each stone was finely cut to interlock with surrounding stones forming load bearing joints without the use of mortars. The stone blocks fit together like a puzzle. The intricate stonework with perfect right angles and smooth surfaces look as though they used machine tools or even lasers. It is located at an altitude of 12800 feet without any natural tree line, and the massive stones were hewn at quarries over 60 miles away. Ancient Alien Theorists have hypothesized that extraterrestrials did help create intricate stonework cut with laser precision into gigantic stones on the top of the desert plateau.

COSTA RICA STONE BALLS: Artificial stone balls, ranging in size from a few centimeters to more than two meters in diameter, and weighing about fifteen tons lie about in the middle of the jungle, and on high mountains, and in river deltas in the small Central American state of Costa Rica. A collection of over two hundred stone spheres has been found in Costa Rica. They are believed to have been carved between 1500 and 500 BC by a civilization long since disappeared, although exact dating is likely impossible. Most of these mysterious balls are made of granite or lava, whereas the heaviest ball weighs eight tons. There is a mystery about the origin and meaning of the stone balls, as they are perfectly executed, absolutely spherical and smoothly polished. There are no quarries for producing the balls anywhere near the sites where they have been found. Neither it is possible to transport nor deposit gigantic stone balls in a primeval forest. Ancient Astronaut Theorists, however, find a direct link between the prehistoric massive balls and various pictures of them on cave walls as well as in reliefs, and propose the visit of aliens on our planet in a ball-shaped spacecraft.

EASTER ISLAND'MOAI': The Polynesian Island of Easter Island is a little plot of Earth, located far away from the coast of Chile in Southern America or any continent or civilization. One of the most remote inhabited islands on Earth, wherein lies one of the world's most famous mysteries. There are 887 giant statues (Moai) of human figures with enormous heads that guard its coastline. These monolithic statues stand 13 feet high and weigh 14 tons, but some are twice as tall and much heavier. The moais were intentionally made with different characteristics. sIt is real mystery to know human beings without sophisticated tools or knowledge of engineering to have crafted and transported such incredible structures. Archaeologists do not know why the statues were built? What they signified? How they were transported? How they were erected? Or Why they were abandoned unfinished? The people of Easter Island themselves are something of a mystery as it remains unclear where they originally came from. Ancient Alien Theorists believe it as the work of visiting extraterrestrials who left their mark on the island.

PERU "NAZCA LINES": Located in South America in the Peruvian spurs of the Andes, lies the ancient city of Nazca. In the high plateau of Nazca, etched are a series of ancient geometrically arranged gigantic lines stretching more than 50 miles. They stretch for miles, sometimes running parallel to each other, sometimes intersecting or joining up to form trapezoids with drawings of animals, birds, and humans. Additionally, there are over 300 geometric designs, which include basic shapes such as triangles, rectangles, and trapezoids, as well as spirals, arrows, zig-zags, and wavy lines. Other examples include a spider, hummingbird, cactus plant, monkey, whale , duck, flower etc.The phenomenon on the pampa of Nazca is a great puzzle as the lines were laid out toward points of compass. Who made the pictures? Why did they make them? How were they able to put each line in its place and align over long distances? Because of their colossal size, the figures can only be appreciated from way up in the air. Ancient Alien Theorists expound that the gigantic lines served as runways for alien craft and the enormous figures guided spaceships as they came down for a landing.

TIAHUANACO "GATE OF THE SUN": In western Bolivia, at a height of 13,000 feet lies the ancient city of Tiahuanaco. It is believed to be one of the most important cities of ancient America. As Andean legends claim the area around lake Titicaca was the cradle of the first humans on Earth. According to the myths Lord Viracocha, the creator of things chose Tiahuanaco as the place of creation. Some researchers Suggest that they date to 14,000 BC. One Ruin still standing in Tiahuanaco is "the Gate of the Sun" carved on a single block of the stone with its imposing frieze of figures. The figures that decorate the stone resemble human-like beings with wings and curled up tails, and wearing rectangular "helmets". The Sun God is at the centre, sculpted with rays emitting from his face in all directions, and believed to have astronomical connotation. Some believe that it was a portal to another dimension or perhaps the land of gods. The Gate of the Sun: A megalithic solid stone structure or ancient monument in Bolivia is confusing experts ever since its discovery. But Ancient Alien Theorists propose extraterrestrial connections to this place.

YUCATAN "THE TEMPLE OF KUKULKAN": El Castillo or the Temple of kukulkan is a Mesoamerican step-Pyramid built around 1000 years ago. It dominates the centre of the Chichen Itza archaeological site in the Mexican state of Yucatan. This pyramid of Kukulkan has a pleasing symmetry with its true majesty lying in the secrets of its construction. The pyramid is a giant calendar. Each of the four faces incorporates a broad steep staircase consisting of 91 steps that ascends to the top platform. The top platform if counted as an additional step, gives a total of 365 steps, number of days in the solar year. During the spring and autumn equinoxes, the shadow cast by the Sun on the northern staircase appears to cause a massively long "Snake" to crawl down the building and link with the stone serpent's head at the foot of the staircase. Ancient Alien Theorists point out that this monument demonstrates a thorough knowledge of astronomy surpassing knowledge of later cultures. They suggest that the ancient Mayans got their advanced knowledge in astronomy, mathematics and other physical sciences possibly from ancient alien visitors.

EGYPT "OBELISK": An obelisk is a tall, four-sided, narrow tapering monument which ends in a pyramid-like shape at the top. Ancient obelisks were often monolithic. Obelisks were prominent in the architecture of the ancient Egyptians, who placed them in pairs at the entrance of temples. An obelisk is said to resemble a petrified ray of the sun-disk. A number of ancient Egyptian obelisks are known to have survived. The earliest temple obelisk still in its original position is the 68-foot and 120 ton red granite obelisk of Senusret I of the twelfth Dynasty at AI-Matarriyya part of Heliopolis. The obelisk symbolized the sun god Ra, and during the brief religious reformation of Akhenaten was said to be petrified ray of the Aten, the sundisc. It was also thought that the god existed within the structure. Now the unfinished obelisk is found partly hewn from its quarry at Aswan. In the ancient quarries near Aswan, Egypt lies a gigantic piece of stone which was intended to be erected as an obelisk. The sheer size of this monument is what makes it remarkable. It is 42 meters tall and weighs 1200 tons. Ancient Theorists purport that it was the work of beings from another planet.

TURKEY "GO BEKLI TEPE": Gobekli Tepe is generally considered to be the oldest religious structures ever found. Some researchers suggest that the structures were built between 10,000 and 9000 BC. The site contains numerous stone structures and stone pillars which feature carvings of various predatory animals. The stone pillars- some of them weigh nearly twenty tons in weight. The site is still mysterious as it dates to a time when humans were thought to be simple hunter-gatherers or the structures were built at a time when humans were thought to be basically cavemen. Archaeologists wonder about the purpose of the site and how did they quarry huge pieces of stone, and cut to size with no metal tools? They are massive carved stones crafted and arranged by prehistoric people who had not yet developed metal tools. Gobekli Tepe is one of the most important archaeological sites in the world. It is perhaps the world's earliest temple predating pottery, metallurgy, the invention of writing, the wheel, and the beginning of agriculture. Ancient Alien Theorists speculate that simply be that unknown beings used this place for their own ends.

EGYPT "GIZA PYRAMID": The great Pyramid of Giza is the oldest and largest of three pyramids in the Giza Necropolis of Egypt. The pyramid was built as a tomb for fourth Dynasty Egyptian Pharaoh Khufu around 2560 BC. The Great Pyramid consists of an estimated 2.3 million lime stone blocks with most believed to have been transported from nearby quarries. The largest stones in the Pyramid, found in the chamber, weigh 25 to 80 tons and were transported from Aswan more than 800 kilometers away. If the perimeter of the pyramid is divided by two times the height, you get a number that is exactly equivalent to the number "PIE" (3.14159). The line of longitude and the line of latitude that the pyramid lies on is 31 degrees north by 31 degrees west. The SPINX with a head of lion matches the belt of Orion as well as the constellation of Leo. Ancient Alien Theorists purport that Egyptian pyramids indicate alien contact and aliens with their plethora of wisdom have definitely constructed the pyramids. The Great Pyramid is still the largest, most precisely built, and most accurately aligned building ever constructed in the world. The Ancient Astronaut idea posits the aliens as builders of pyramids, and so forth.

Ancient Alien Theorists have also noted aliens as the secular equivalent of angels and demons and ghostly spirits. Ancient Alien Theory reasons that alien globe-spanning engineering projects were far beyond our own capabilities. Many UFO researchers have found correlations between the pyramids of Egypt and what seem to be pyramids on Mars, the Moon, and Venus. Not surprisingly, from that has come the suggestion that perhaps the pyramid design antedates the Egyptians, who did not invent the structure but apparently followed an already existing model. Questions abound not only regarding the origin of the Great Pyramid but of the structure's purpose and the intent of its builders. The pyramid forms are found not only in Egypt but also in Mexico and Central America. They sometimes appear to adopt a theory of aliens utterly contrary to that of mainstream science. Scientific resistance to such revolutionary ideas results from the fact that science does not accept radical ideas quickly. It is good for our soul, and for our intellect, and good for our work to have our imagination stretched, to be open to the possibilities.

CHAPTER SIX

Pyramid Technology

Evidence is overwhelming that the Earth is being visited by intelligently controlled vehicles from off the Earth-Stanton Friedman

Pyramids are tangible enigmas, ancient remnants of a time beyond memory, beyond history, and beyond understanding. The colossal architectural edifices have, over the centuries, provided archeologists, historians and mystics with material for numberless theories, endless debates, and inner meditations. Today, the mysteries of pyramids still continue to intrigue and plague scientists, scholars, and all those researchers. In fact, there are ancient pyramids all around the world, from a number of different cultures and civilizations with many different architectural styles and approaches. It is not surprising to learn therefore the list of world pyramids really does span the globe. From the famous pyramids of Egypt, Mesoamerican pyramids, Chinese tomb pyramids, south American step pyramids, Mesopotamian Ziggurats, North American mound pyramids, and to even Roman ceremonial pyramids: these ancient structures pop up across the globe right through the centuries in cultures who often have no connection to one-another. No matter if the civilization was Mesopotamian, Egyptian, or Mayan, its legacy today is in part marked by towering pyramids.

We can only speculate as to their purpose or how ancient man roughly five thousand years in the past had the technical prowess to accomplish such a feat. In addition, we find that almost all of the pyramids around the world are designed with some type of astronomical alignment in mind and likely very advanced sky observations alone just to be able to infer configurations. During the past hundred years, pyramids have been recorded with varying degrees of accuracy as to location. Most of these pyramids have been sighted by military pilots flying over uncharted areas during their flight missions. A white pyramid has been located somewhere in the Himalayan Mountains, and is described as shimmering white. Encased in metal or some sort of stone, with a huge capstone made of a jewel like material, possibly a crystal. A big complex of pyramid structures with one large pyramid is apparently located in the Shensi province of China. The Shensi pyramids seem to be constructed from a mixture of lime and clay, hardened into a cement like material. They are covered with casing stones and decoratively painted in various colours.

In the jungles of Cambodia lie the ancient ruins of the once great city, now known as Angkor Wat, which contain splendid temples, endless galleries and vast pyramids. A complex of pyramids is reported to have existed in a desert region in the central Siberian uplands, north of Olekminsk. A pyramid-like structure is found in the south of France, beneath which is a subterranean pit with astrological symbols carved into the walls. Silbury Hill, in Willshire, England is one of many British cone-shaped mounds or stepped earthen pyramids. The Castillo in Peru is an immensely complex building, seemingly pre-pyramidal in structure. There is a great deal of mystery surrounding pyramids- from the enigma of the building to the colossal Egyptians, Mayans, and Peruvian pyramids to the perplexing and inexplicable powers seemingly intrinsic to the pyramidal shape. Of course, the first mystery of the pyramids is that of the origin of the name itself. The controversy over the derivation of the word pyramid is minor, compared to that which rages over the purpose of the pyramids themselves. Archeologists investigating pyramids claim that Egyptian pyramids were used as temples.

But some experts now believe that the pyramids are, possibly, resonators or storehouses of energy. Ancient and mysterious locations of pyramids can be connected along the same belt, around the world between 20 degrees and 32 degrees north latitude. If you take the line of longitude that the pyramid lies on and the latitude that the pyramid lies on 31 degrees north by 31 degrees west, they are the two lines that cover the most combined land area in the world. In essence, the pyramid is the centre of all of the land mass of the whole earth. The Belt also connects the Bermuda Triangle, the Dragon's Triangle tending to be an epicenter for anomalies and geomagnetic fluctuations according to some theorists. The fact that the Egyptians had not even invented the wheel yet, but the blocks that they had to carry to build the pyramids weighed about 2 tons each, and then how did they transport them? The question who laboured to build them and why, has long been part of their fascination. The Great Pyramid of Egypt is more profound, and awe inspiring: two million, five hundred thousand stones, with crushing weights of two to seventy tons, rise to a height of more than four-hundred and eighty feet.

This awesome structure, by is sheer bulk alone, staggers the imagination of modern construction engineer. The precision with which the stone was cut and positioned indicates that the builders of the Great Pyramid were masters of measurement. The blocks from the Great Pyramid of Giza are cut so precise that even a sheet of paper cannot fit between two of the blocks. The three pyramids of Giza also are a perfect reproduction of the three stars of Orion's belt. In addition, it has been recently discovered that the air shafts of the great pyramid project directly toward the stars in Orion's belt: apparently with the aim of projecting the soul inside the pyramid directly into heaven. They practiced highly advanced techniques in the art of mummification. Drawings and sagas actually Indicated that the gods promised to return from the stars in order to awaken the well-preserved bodies to new life. The Pharaoh certainly knew more about that the gods who would return and wake him up. In Mexico on the winter and spring solstice, the Mayan pyramid at Kukulcan casts the shadow of a giant feathered snake that slithers down the steps to reconnect with its disembodied head at the bottom of the pyramid.

This happens every March and December 21st as the Sun rises in the sky and hits the edges of the pyramid's steps. Legend says the snake is representative of the Mayan king-Kukulcan, who arrived to rule Chichen Itza on a bed of snakes in the 10th century. It is also believed that the snake represents the kundalini serpent, which is symbolic in Indian philosophy for the source field energy, entering back into the Earth from Heaven, where it originates. This is probably both a mythic interpretation and scientific visualization of what is actually occurring within the pyramid. The pyramid of the Sun in the sacred city of Teotihuacan, Mexico gives the illusion of infinite height and space. This Mayan pyramid is so designed that an observer standing at the base of the great staircase cannot see people at the top. These pyramids give us a new sense of wonder and mystery about the remarkable structure and its effect on humankind. So the logical questions that naturally follow are- How was ancient man able to construct such vast monuments? Who built the pyramids? What was the purpose? Where from the unknown builders acquire the extraordinarily advanced scientific and astronomical knowledge?

These questions while still unanswered, continue to pique the imaginations and curiosities of many scholars. The problem is that whoever built these structures did not leave any evidence of the tools they used to do so behind, perhaps this was done on purpose. Compared to our construction today, the pyramid construction technology was far superior. It appears that the only way our ancestors could have been able to construct such megalithics was with the aid of an advanced civilization- perhaps thousands or millions of years ahead of our own. The technology needed to construct these monuments was far beyond anything the ancients were supposed to have at the time. There is no doubt that every civilization which built pyramids did so with the use of highly advanced mathematical and astronomical calculations and seemingly impossible mastery of the skill of stone masonry. So, it would appear that some type of advanced civilization at least had a hand in their construction. In civilizations separated not only by thousands of miles, but by hundreds of years, stones weighing many tons were hoisted into position with infinite precision for the purpose of erecting pyramidal structures.

Also our ancestors, who were separated by vast oceans, had no knowledge of each other's existence, but were somehow constructed strikingly similar structures all over the world for a reason that is still a total mystery to us. The speculation is ripe that the pyramids were constructed by other beings from elsewhere in the galaxy. Because of the virtually identical use of skill and science in the erection of the pyramids, it is impossible not to conjecture that perhaps these skills and science were taught to pyramid builders by persons from outside the civilizations. These extraterrestrial beings millions of years ahead of our own might have seeded life on this planet. Possibly these pyramids have been built by this older generation as a sign of our connection to the stars and as a prerequisite for rejoining our galactic family. Of course, we are all an integral part of the cosmic evolutionary process. Perhaps the ancients used this stone as a type of technology, which may be unlocked in the future. Now, we use today, the silicon quartz stone crystals to power every microchip - from cell phones to laptops to ipods. The pyramids in Mexico and Egypt were built by the cultures separated by a vast ocean without any contact.

The deeper we dive into this mystery, the more clear a set pattern emerges. There is the need to place monuments such as the pyramids, the stonehenge, Teotihuacan, and other incredible places in specific locations, creating a pattern that we today identify and connect. Indeed, they were not placed randomly. The pyramids, however, serve other less known or understood purposes. One of the more interesting uses for a pyramid is that of an energy balancing mechanism for the Earth. All the pyramids around the world seems to have been built on certain 'nodes' or lay line positions. The pyramids appear to connect to points on this grid and are part of a network that is significant to the functionality of the Earth. The pyramid structure itself acts like a funnel for hyper dimensional 'source field' energy pushing down into the Earth from space as the force we call gravity. The energy is quite literally the divine force: God, which creates the entire universe from the state of pure, blissful formlessness to what we experience as the physical universe: from the state of pure Being to Becoming. Physics today tells us that gravity is the least understood force in physics.

Once we master gravity, it will open to a brand new type of physics. Our ability to manipulate, control and create artificial gravity will become commonplace. Michio Kaku, the famous theoretical physicist says the civilization that has mastered gravity is able to create limitless, pollution free energy and manipulate the basic fabric of space-time to traverse vast interstellar distances and capable of even travel through time. The pyramid technology may help us learn more about this new type of physics and 'source field'. The source field is the sum total of all matter, energy, space, time, biology, and consciousness. A new understanding of gravity recognizes that it is not something that pulls us down toward the Earth, but rather something that pushes down on us from above. This energy is what we call gravity but in reality is the God-like force that creates and maintains the universe itself. This new understanding of gravity links it more to a fluid-like energy moving into the Earth from the very fabric of the space-time universe itself. A pyramid seems to be a type of technology that utilizes this force by harnessing the source field energy of gravity and concentrating it through the geometry of the pyramid.

This may be more common knowledge in main stream Russia, where recently they have been building fibre-glass pyramids all over their country as research into 'source field' physics. They appear to have documented many benefits of pyramids which harness this galactic energy field. The pyramids seem to harmonize their surroundings. We do not know whether these pyramids were intentionally created as a message, but only time will tell when we decode their true meaning. Today mainstream researchers believe that embedded in these megalithic constructions hides a secret code that can explain how, why, and who built and organized these monuments. If so, from where did these outsiders come? How did they get here? What was the motive behind the endowment of this knowledge upon the ancient civilizations? Was there a purpose to evolution that our ancestors understood and encoded in these pyramids? What type of ancient technology that our ancestors used in ways we never could have imagined? What about the destiny of the human race and what can these pyramids all around the world tell us about it?

Ancient astronaut theorists believe that aliens, with their plethora of wisdom, came down around 10000 BC and built the megalithic monuments. UFO Aliens may have helped build mysterious pyramids around the world. There seems something inside the pyramid that is not of this world. Ancient Alien theorists further state that the ancient flight patterns appear to be arranged in straight lines along the ancient monuments and megalithic structures. These lay lines are magnetic in nature and they delineate the lines of Earth's unseen energy fields. Another interesting fact is that all of these locations are still associated with a large number of UFO sightings. These visitors seem to understand the importance of these places and frequently visit them and obtain energy from them. Ancient astronaut theorists believe our ancestors purposely constructed their monuments on energy lines that when mapped and connected create a significant pattern. Ancient civilizations surprisingly knew that if they placed their monuments in specific locations, they believed that everything would come into flow and into place. Ancient Alien theorists say that our ancestors purposely constructed their monuments on energy lines.

CHAPTER SEVEN

Alien Search

"I am sure the universe is full of intelligent life: it has just been too intelligent to come here"
-Arthur C. Clarke

One of the greatest mysteries of the universe is the mystery of aliens and UFOs. People wonder if they do exist, what are they? Where do they come from? What is most important of all is what are they here for? SETI is an acronym for the Search for Extraterrestrial Intelligence. SETI is a scientific effort to discover intelligent life elsewhere in the universe, primarily by attempting to discover radio signals that indicate intelligence. Although SETI scientists publicly proclaim their disbelief in UFOs, they nevertheless believe that intelligent extraterrestrial life does exit. SETI looks for radio waves in certain frequencies for signals and patterns that could prove the existence of extraterrestrial life especially in space. This is done with the use of large radio telescope. There is electromagnetic radiation in the universe, so it is natural for an extraterrestrial intelligence to manipulate this to try for contact. Search for radio signals or optical flashes from other star systems that would indicate the presence of extraterrestrial intelligence have for over fifty years so far proved fruitless, but detection of such signals would have an enormous scientific and cultural impact.

Current methods of SETI are very anthropocentric. We listen only to frequencies we think important and these are based on our body chemistry. We are ignorant about in which frequencies the aliens are transmitting messages and we would never dream of examining them. Also radio frequency interference poses one of the most significant technological challenges to any ground-based SETI program. In 1961, Frank Drake, a radio-astronomer, built a simple formula, known as Drake Equation to determine the number of detectable civilizations in the Milky Way galaxy. This number is equal to a series of factors, each of equal importance, simply multiplied together. Start with the rate at which stars are formed in the galaxy and multiply this by the fraction of stars that have planetary systems. Then multiply by the fraction of those planets hospitable to life and by the fraction of those planets on which life has actually developed. Multiply by the fraction of living planets that evolve intelligent life forms and by the fraction of intelligent life forms that want to communicate and are capable of doing so. Finally multiply by the length of time that such a communicating civilization remains detectable.

When all of these variables are multiplied, N equals the number of communicating civilizations in the galaxy. The aliens are mortal creatures that can dwell within the atmosphere among planets: hey have mortal material needs, reproduce through DNA manipulation, and eventually will die. Science fiction has often depicted extraterrestrial life with humanoid or reptilian forms. Aliens have often been depicted as having light green or gray skin, with a large head, as well as four limbs-fundamentally humanoid. However, the UFO fraternity believes there are many different species of alien visiting

Earth, all of whom interact with human beings. Many researchers for varying reasons have come to believe that there are multiple species of extraterrestrials. The most common aliens that are involved among humans are the Greys. The Greys are foremost in our minds when we think of aliens. They are referred to by some as Zeta Reticulans, believed to be the culprits behind the abduction phenomenon, which has earned them their nefarious reputation. Among the different races of aliens are small greys, large greys, reptilians, and Nordics.

Some of these alien races are purely benevolent, contactees say, while others are ultimately hostile. Small Greys seem to possess an electronically monitored and controlled social memory complex that allows them to function effectively in a group-mind mode. Tall Greys seem to have technological superiority, but they appear to lack in spiritual and social sciences. There are alien beings called Traders: androgynous beings who have travelled the universe for billions of years and long ago discovered a way to channel energy from black holes to supply their planet. Zetas are the small greys most commonly identified as aliens or ETs, and come from the star system Zeta Reticuli. Zetas do not communicate via spoken language but instead transmit their thoughts telepathically. The Reptilian race is higher ranked than the Greys. Conspiracy theories abound that the Reptilian is a malevolent group who seek dominion over not only our world but others all around the universe. Those who believe in this theory, claim that the Reptilians live in third, fourth, and fifth universal dimensions and feed on negative energy. The Reptilians and the Grays are closely involved in human affairs.

The Reptilians are six-foot-tall race of creatures, as contactees say, who have actually contributed DNA strands to our genetic makeup. These beings are explorers, mapping out distant parts of the universe for their Trader superiors. The Reptilians have taught their contactees that the entire planet Earth is alive and that our distinctions between animal, plant, and mineral are based on completely false assumptions. In fact, they have said, every object on our planet is alive and to a degree sentient in its own way. The Reptilians seem to carry the messages of preservation of earth's resources. One of the most talked about subcategories to the Reptilian race is the Draconians or Dracos. Many believe that the Reptilians possess shape-shifting abilities. Nagas are another type of Reptilian being. According to Hindu legends they are linked to snake world, a multi-levelled cavern system under the southwestern slopes of the Himalayas which is the home of Nagas. The Nordics are tall, blond, and almost indistinguishable from human beings. This race of beings as per contactees, act as caretakers, and their members are quite benevolent.

Many Contactees do not even realize that the Nordics as extraterrestrials because of their human-like appearance and believe them to be angelic creatures. Nordics are believed to have the agenda to enlighten by providing spiritual revelation and warn humans about behaviours that could potentially lead to catastrophe. Some proponents feel that ideas within the New Age movement were given to us by the Nordics. The Mantis beings appear as six-foot-tall creatures that resemble a praying mantis. They have a specific role in the contact/abduction experience paralyzing their subjects with an electrical pulse to prevent them from injuring themselves because of the human fright or flight reaction. In this capacity, the Mantis beings act almost like anesthesiologists during the contact experience. Aliens are higher dimensional beings. The realm that aliens function in is a dimension in reality that is above us. Aliens are very much in tune with psychic level of reality. Alien as inter-dimensional being is a very complex subject. No scientific perspective of aliens can be complete without taking into account the spiritual aspects of it. Higher dimensional reality is spiritual and mental in nature.

Many have referred to the higher dimension as the mystical plane. With the awareness of our soul, we can often view our highest spiritual lives and perceive our multidimensional Self. Physical science alone cannot explain such things because they are of a higher dimension than the physical world. Both the quantum physics and metaphysics are needed for the full understanding of the higher dimensions. The true nature of reality is that everything is made of pure consciousness. Physical reality is only a condensed form of consciousness. Aliens as well as UFOs move in and out of physical reality. They are not just figments of our imagination but they possess spiritual existence. Aliens are hyper- dimensional beings: more dimensions mean more freedom of movement. Our physical reality that we experience has three dimensions namely length, width, and height. There is then a fourth dimension called time, in which we experience one single point in time at a time. In reality all time is happening now. We cannot conceptualize the fourth spatial dimension, so its inhabitants would seem omnipotent to us. Beings from higher dimensions can see us, and can physically move down to our level.

But we cannot move up until we are able to readjust our frequency levels to one that is infinitely finer. One of the reasons why we third dimensional beings cannot see other life forms, even though they are all around us, because we are not ourselves at a high enough energetic frequency level to do so. That means the three dimensional life forms physically operate at a lower frequency of vibrations than those of higher dimensions. The dimension that is above physical reality vibrates at a much higher frequency. It also possesses a more thought like state of reality. Beings from the fourth dimension would be able to walk through walls, reach through solid barriers, appear or disappear on a whim, and materialize in whatever location they please. They hide behind the veil of perceptions and can project themselves into our visible reality at will. It seems certain that our bodies have difficulty living multiple points in time simultaneously. What happens beyond fourth dimension may have something to do with spirituality. Lower dimensional consciousness is based on a materialistic viewpoint of life and physical matter is very dense.

Quantum physics reveals us that the phenomenon of consciousness involves subatomic particles; it tells us that telepathy can hypothetically occur instantaneously between two individuals which are light years apart on different worlds. We seek to explore the concept of quantum realities in which parallel universes and multiple dimensions are real. It seems reasonable to assume that UFO occupants know more about quantum physics than we do. Implanted visions would probably within their ability, either through technology or sheer mind power. UFO occupants appear to have achieved a deep reservoir of implanted visions within the human mass consciousness; they are a part of our own mass consciousness. There exist positively oriented beings with the agenda of assisting the ascension of planet Earth. The alien-angels lend their services to help humanity evolve spiritually. Their constant interactions with humans are to open up our minds to the worlds beyond our own. These alien beings guide those of us who seek freedom from the limitations of the 3-D matrix control system. Without intense thought and concentration, we three-dimensional beings cannot visualize a four-dimensional reality.

We have very strong concept of death and destruction in our three-dimensional reality with a feeling that there will soon be an end to all that is. Biocentrism experts claim that the death as we know it cannot exist in any real sense. We have been taught to accept the idea of dying but in reality it just exists in our minds. The theory of biocentrism explains that life is an adventure that transcends our ordinary linear way of thinking and the evidence lies in the idea that the concept of death is a mere figment of our consciousness. The angelic aliens seek to raise collective consciousness towards mass enlightenment so that the earth vibrations will become higher dimensional in nature. The raising of our vibration frequencies leads to the escalation of our conscious awareness of higher energies and higher dimensions. We tend to evolve into beings who live life from higher levels of consciousness with the awareness of this dimensional change. The gateway to these planes lies within us by looking into the knowledge of metaphysics. Higher planes of reality can be perceived only by our inner senses. Our observable universe began at a finite time in the past in a hot explosion- the big bang.

The fading glow of the ultra-hot expansion brought into being space and time, matter and energy. The evolving logic of the cosmos defies our imagination. The universe assumes its familiar appearance of darkness punctuated by islands of starlight. In charting some of the billions of galaxies, we have been stunned by the immensity of the universe. We are spreading the light of reason out across the universe, which will teach us how to look at things in a different way. The universe has to be large for life to have evolved. True, there are far more energetic phenomena out there. The very materials out of which our bodies and of aliens are made were cooked up in a violent process somewhere in the universe. If life is channeled by the forces of nature and evolution, life forms somewhere or other in the universe need to be very similar to ourselves. Only when we contact a form off the Earth, we will know which parts are universal in the truest sense of the word. There is a creative intelligence that governs the universe. There is a connecting intelligence linking all things, big and small. Alien beings have been around for billions of years, so long it is almost impossible for us to comprehend.

These alien beings have been in existence long enough to observe the universe and know there is order to all things. However, the discovery of extraterrestrial life would not contradict basic conceptions of God. The science of aliens is getting weirder still. If we were the creation of aliens, then who created aliens? – It is the universal Divine Spark of Cosmic Intelligence. We need to be far more expansive in our efforts, by questioning existing ideas of what form an alien intelligence might take, how it might try to communicate with us, how we should respond if we ever do make contact. Our research need to focus on the big questions of existence ranging from the origin of the universe to the origin of life and the nature of time. At this point, we inevitably leave the domain of evidence-based physics and enter the world of metaphysics. Spiritual unfoldment consists in coming to realize, that we humans and aliens are the same life that animates all else. There is an energy source that runs throughout the universe and flows through all dimensions. We are enmeshed in the universal processes of Nature and are in essence part of the One Source of all. Life is one, but it has infinite aspects and pervades endless dimensions.

CHAPTER EIGHT

UFO Phenomenon

"I am convinced that UFO have an out-of-world basis"--- Dr. Walther Riedel

UFO is an acronym for Unidentified Flying Object. In common language UFO has been often synonym for an alien spacecraft. Culturally the phenomenon has often been associated with extraterrestrial life and has become a popular theme in science fiction. UFOs have become a relevant theme in modern culture and the social phenomena have been the subject of academic research in sociology and psychology. These objects appear to be technological but not natural phenomena, and are alleged to display flight characteristics or have shapes seemingly unknown to conventional technology: the conclusion is they must not be from Earth. UFOs constitute a widespread international cultural phenomenon of last 60 years. UFO may be stated as any airborne object which by performance, or aerodynamic characteristics, or unusual features, does not conform to any presently known aircraft or missile type, or which cannot be positively identified as a familiar object. Though UFO sightings have occurred throughout recorded history, modern interest in them dates from the Second World War. A number of military personnel and others have given statements about having witnessed UFOs themselves.

Thousands of people make reports yearly of strange lights performing impossible maneuvers that traditional earth-bound aircraft, commercial or military could never accomplish. While most of the world's governments try to deny the reality of their existence, they blame the sightings on natural aerial phenomena, swamp gases, weather balloons, or mass hysteria. They have established that the majority of UFO observations are misidentified conventional objects or natural phenomena with a small percentage even being hoaxes. But there are thousands of documented cases of unexplained ships, hundreds of cases of governments holding what they know. One of the reasons for covering up by government agencies of the world is that they would not disclose to their populations that there is a force against which even the most powerful countries cannot mount a defence. The cover-ups and secrecy by governments are largely because UFOs are a sensitive issue. They are forced to discreet silence for inevitable implications of UFOs detrimental to governments. Those government scientists who evaluate the UFO phenomena go through changes as they read the evidence.

They evaluate in such a way that sometimes it is all rather insubstantial, whereas at other times it looks to be very probable. Even the most basic issues provoke controversy with inflated egos in the barbed discussions of the UFO debate. The sighting of strange things in the sky may actually predate the emergence of modern man. But the theme of human-alien interaction was thrust into spotlight some 60 years ago. If you have seen some strange light in the sky, indistinct by unmistakably moving in a way that denies ordinary explanation, then you may have seen UFO. If you have seen some strange thing in the sky that you could not explain and that would therefore qualify as an unidentified flying object. We need to go beyond UFOs and beyond the arguments based on personal beliefs and opinions and turn to science. Ufology is a neologism describing the collective efforts of those who study reports and associated evidence of UFOs. Ufology comprises credible witnesses seeing incredible things. Ufology has grounding in hard facts, and there is no doubt that ufology adds a thrill to life. Any sensible ufologist would dispute misperceptions of mundane phenomena or mistaken ordinary phenomena.

It is wrong to dismiss that ufology or the UFO field as the province of egotistical dreamers with delusions of grandeur. There are many well-qualified researchers and highly intelligent scholars from engineers, geologists, physicists, psychologists, and meteorologists who have been drawn by evidence, not by dreams. Many UFO encounters provide evidence that aliens have been visiting Earth in UFOs is true. Hard evidences such as photographs or pieces of the UFO sightings have been presented. There are millions of witnesses from almost every nation on the planet who claim to have seen something inexplicable in the sky. Many of them are indisputably sincere, rational, lucid, and intelligent; occasionally they are even astronomers, scientists, military personnel, airline pilots, police officers, and skilled observers. Yet nobody seems to know for certain whether UFOs prove the reality of spaceships. Above all else, the evidence may be suggestive but do not prove that to the satisfaction of science. Most pundits of ufology embrace far more readily the extraterrestrial hypothesis and establish it as the best solution on offer for the evidence.

They dispel the myth of para-physical creation of the mind, or psycho-social perspective, or the stuff of fantasy etc as incredibly thin edifice of theory. Ufology is now regarded as potentially among the most important subjects in the world. Many ufologists argue that people all over the world are reporting essentially similar encounters with aliens and UFOs, and the overall emphasis is widely in favour of true alien presence. The UFOs and alien visitations now convince and conclude not just that other civilizations are possible, but that they really exist. As UFO sightings appear often, and in substantial numbers, they might have far more advanced technology. The report we hear of UFOs visitation, the smart aliens have to come from far away stars. The alien technology would have to be far more advanced to travel quite easily among the stars. Their advanced technology allows them to do things like cloaking their spacecraft to prevent us from seeing them. It seems possible that alien races pretty much focus on faster-than-light travel, energy weapons, and artificial gravity. For some UFOs are a status symbol, since they can elevate their respectability in an unusual sort of way.

The time travel, remote viewing, out-of-body projection, and psycho-kinesis have become the collective holy grail of exotic science of UFO. If you believe in UFOs or have seen one, makes you the centre of attention. Some people tend to favour the view that UFOs might be a manifestation of a sort of dimensional interaction. There are the debunkers who usually cry' hoax', often with doubtful credibility. The UFO evidence must either reflect real spaceship or some wild fantasy. Physicists liken the UFO to the dilemma of the nature of light: in some experiments it behaves like solid particles and in others it is electromagnetic radiation. If one is stuck with the question are they this or are they that, it is the question that is wrong. The reason is quite simple: our knowledge is not yet sufficiently advanced. Many serious ufologists have a particularly strong belief that those aliens somehow have been here, right on Earth: in the multidimensional richness of human experience on this planet. There are certain cases whose reality of events leave little alternative but to say that they are incredible machines of non-earthly design which have intruded into our airspace as proof of alien visitations.

The aliens may not speak to that part of our consciousness the one we deem most important – our spirit. However, UFOs have been present throughout history and take on an image that is culturally relevant. There is one ultimate truth the world is afraid to accept—the reality that we are not alone in the universe, the proof is here, the proof is real, seek out the truth: Aliens the truth. The truth about UFOs may be that there is no truth until extraterrestrials themselves elect to disclose their own existence. The world should wake up to hear the awesome news: The aliens are here. The early days of space programs were shrouded in military secrecy. There were intense speculations especially among the UFO fraternity: "Cosmonauts saw UFOs" and "Astronauts forced down by UFOs". Ufologists are optimistic that there are things out there. There absolutely are! UFOs are real. Military and commercial pilots have reported seeing strange objects flying near them, but they have been told in no uncertain terms not to file official incident accounts. It is hard to turn a blind eye, or deaf ear to what people have seen and said.

UFO magazines and other media have been covering about sightings of unidentified flying objects, flying triangles, alien abductions, time travel, exotic science, and government disinformation as mainstream stories: the most incredible reports and eye-witness accounts of our time. They seek to take a hard scientific approach to some of the most astonishing UFO sightings reported from around the world, which persists to this day. The activities have been wide spread and more frequent that people would think over the last 60 years. But understanding this phenomenon and why this is happening is most crucial to the whole of mankind. ROSEWELL UFO INCIDENT 1947: An airborne object crashed on a ranch near Rosewell, New Mexico on July 7, 1947. The most famous explanation of what occurred was the crash of a spacecraft containing extraterrestrial life. The Rosewell incident has been the subject of much controversy and conspiracy theories have arisen about the event, even though the crash was attributed to a US military surveillance balloon by the US government. In contrast many UFO proponents maintain that there is a massive cover-up by the military.

CANARY ISLANDS YELLOWISH LIGHT 1976: Thousands of people have seen a spectacular luminous phenomenon in Canary Islands on June 22, 1976. Eyewitnesses included civilians, physicians, priests, and engineers, who saw an intense yellowish-blue light moving out from the shore towards their position. When it became stationary, the original light went out and a luminous beam from it began to rotate. After some time the upper part began to climb in a spiral, its glow lighting up the land and the ocean. RUSSIA STRANGE CRAFT 1989: Several children from the city of Voronezh, three hundred miles south of Moscow, saw a strange craft land nearby from which emerged two large beings and a small "mechanical man" in the year 1989. The children were playing at the city park in the early evening when they witnessed what has been variously described as a "Pink, shining object in the sky and as a red-coloured ball that measured around thirty feet in diameter". The children as well as the crowd could clearly see a hatch opening in the lower part of the ball and a humanoid in the opening. A three-to- four-meter-tall, small-headed, three-eyed being- wearing silvery overalls and bronze boots- was seen inside the sphere.

BELGIUM TRIANGLES 1989: The triangular- shaped craft sighted repeatedly over Belgium from November 1989 through April 1990. It was estimated that almost nine out of every ten people in Belgium saw the flying triangles during this period as both day and night sightings. Belgium became the first and only nation in the world to openly and officially recognize the existence of UFOs. PHOENIX LIGHTS 1997: The citizens watched a series of lights- along with a very large tri-angular shaped saucer hovering silently in the night sky over the greater Phoenix, Arizona area on March 13, 1997. The Phoenix Lights, which floated over the desert city before fading out, were one of the most dramatic and important mass UFO events of recent times. Everyone in Phoenix had videotape of the mysterious lights to show the media. Though their origin has never been fully explained, to some they were flares dropped from military aircraft as part of an exercise, but to many investigators they were UFOs, plain and simple. According to the Mutual UFO Network thousands of people reported seeing strange, bright lights flying over Nevada, Arizona, and part of Mexico.

CHINA UFOS: A sudden proliferation of sightings in China began since 2, December 1999, causing enough of a sensation. People in Shanghai observed a shining cylinder with a flaming orange tail scooting above the skyscrapers, almost every city in China. Other cities reported glowing streaks on the horizon and the Internet sites began posting blurred photographic evidence. Poor farmers in Beijing's barren hills saw an object swathed in coloured light that some say must have been a UFO. There have been more reports of close encounters with aliens in the Chinese media. UK UFOS: UK is perhaps the place where the number of UFO sightings is very high. Reports continue to fill in the media as many witnesses conferred their respective sightings. These unidentified flying objects were spotted near several major landmarks in the country, including Black-pool Pier, the houses of parliament, and the famous Stonehenge. The National Archives released files which included notable sightings reported to the Ministry of Defence, UFO Desk dating from 2007 to 2009, when such number was high. An Airbus a320 pilot encounters an object closing in on his passenger plane on 13 July 2013 In UK.

INDIA UFOS 2013: Indian Army troops have sighted mysterious UFOs in the Ladakh sector along the Line of Actual Control with China. The UFO was sighted by Indian Army troops in Lagan Khel area in Demchok in Ladakh area in the evening on 4 August 2013. The reports suggested that yellowish spheres appear to lift off from the horizon on the Chinese side and slowly traverse the sky for three to four hours before disappearance. In October 2015 a UFO sighting was reported in the region of Ratnagiri district in Maharashtra by 800 people from villages. JAPAN UFO 2019: Fuji News Network reported capturing footage of illuminating object in the sky traveling at tremendous speed in Iwaki City, Fukushima prefecture. A mysterious balloon-like object was seen floating across the skies of northern Japan on 18 June 2020. A close-up image suggested that the object was equipped with solar panels. The world UFO day is celebrated on every year on 2 July-the day when the supposed Roswell UFO crash took place in 1947. It is an awareness day celebrated by UFO enthusiasts with the motive to educate people about the undisputed evidence that UFOs are real.

BERMUDA TRIANGLE: The Bermuda Triangle has long been thought to be a realm of intense UFO activity- a section of the Atlantic Ocean bounded at its points by Miami, Bermuda, and Puerto Rica- is a region where ships and planes have mysteriously vanished without a trace. Unexplained circumstances surround some of these accidents, including one in which the pilots of a squadron of U.S. Navy bombers became disoriented while flying over the area; the planes were never found. Other boats and planes have seemingly vanished from the area in good weather without even radioing distress messages. Along with Unidentified Flying object and Unidentified Submersible Object sightings, there have been reports over the years of bizarre magnetic and atmospheric phenomena that render ship and aircraft compasses and engines useless. Marine geologists have studied the seabed beneath the "Triangle" and found it to be geo-thermally active. The gas envelop spread over huge area of ocean surface, and into the atmosphere is believed powerful enough to cause passing ships to sink, and combustible enough to explode engines.

CROP CIRCLES: Crop circles have become associated with UFOs because they often appear after UFO sightings. There is the assumption, which is not necessarily true, that the cause must be extraterrestrial, as no one on earth can duplicate the energy process that imposes these designs on the stalks of grain. Elaborate designs have been formed and discovered in England, Australia, Europe, Japan, Canada, and the United States. Within these mysterious circles, the stalks of grain have been laid flat, usually in a circular or flowing pattern. Crop circles first began appearing in grain fields in England in 1975 and the designs of these circles became increasingly more complex since 1990. Many of them cover huge field space and their crop circle formations are absolutely stunning in their geometrical design, intricacy, and precise measurement. Crop circles have a strong magnetic field in and around them. They all exhibit an electromagnetic resonance that extends from the centre of the crop circle design outward in diminishing order. They correspond to mathematical equations, sacred geometry, electrical circuitry, musical notes, DNA strands, asteroid trajectories, solar system, and hyper dimensional physics.

CHAPTER NINE

UFO Encounter

"To my way of thinking, there is every bit as much evidence for the existence of UFOs as there is for the existence of God"- George Carlin

One of the world's greatest mysteries: are alien life forms visiting Earth and performing experiments upon unwitting human victims. There are UFO stories that originate from almost every country in the world. The subject of alien abduction is extremely complex. Experts involved in ufology have a collection of true stories of people who consciously saw them as physical beings. There are numerous stories of fantastic craft that can do incredible feats, which offer the only solution that they have to be alien space ships of some kind. UFOs are usually small flying objects: glowing orbs, metallic spheres, satellite-like flying machines etc. Normally their flying patterns suggest that they are not of the world- the science of extraterrestrials. The terms alien abduction or abduction phenomenon describes subjectively real memories of being taken secretly against one's will by apparently by nonhuman entities and subjected to complex physical and psychological procedures. People claiming to have been abducted are usually called abductees. Due to paucity of objective physical evidence, most scientists and mental health professionals dismiss the phenomenon as deception-hypnosis, fantasy-proneness, or sleep paralysis.

Ufologists focus specially on sightings and try to eliminate as many natural causes as possible. UFO researchers try to make rigorous examination of all relevant facts to find a natural explanation for the event. Only those sightings that can withstand such vigorous tests deserve to be investigated further. For some ufologists in addition to the big secret of UFOs themselves, it is the process of time travel and remote viewing that fascinate them. There are ufologists who focus on a history of records, and try to come to some understanding of the UFO phenomena. In the field of alien-abduction studies, they make investigation studies into the contact between human beings and the creatures who abduct them. UFO researchers, authors, and people claim to have secret information. Alien kidnapping experiences have similarities in stories, feelings, sensations and spiritual experience. But science believes that such strange phenomena are bought into the story because they are completely weird. Alien abductors have been the subject of conspiracy theories and science fiction stories that alien implants could be a possible form of physical evidence.

UFO luminous phenomena and abduction might be a mystical experience by applying complex electromagnetic field pattern, as brain is an electromagnetic organ. Through brain stimulation it is easy to create false memories. Abduction activities may be major electrical effects which are hypersensitive to electromagnetic field- all types of sensations, dream-like experiences or hallucinations. They seem to work on human consciousness-hypnosis. Alien creatures although non-violent, possess the extraordinary power of reducing the individual's will to resist. Night-time visitation experience and close encounter with flying discs, strange bright shining lights and robotic like beings not human seem to be a great mystery. Aliens land by some kind of light beam from spacecraft. People explain that bluish energy light makes them momentarily blind, but regain consciousness slowly. They often see creatures like robots not humans in close encounters. They often report that they find alien implant in body after they come closer. The implant in body seems to be a metal of non-terrestrial isotopic nature. Again they say aliens have left strange marks on their body, which fade away later on.

But such physical evidences are not conclusive for science. All these encounters and abduction phenomena are likely to prove the real existence of life beyond us or false memories by brain stimulation. It is quite important to know about aliens as we humans and extraterrestrials are made up of the same star stuff. Alien abductions seem to follow a prototypical scenario in which the individual, or small group of individuals, are isolated, paralyzed, and then, while completely conscious, physically transported aboard a spacecraft by some kind of light beam. The abductee often finds himself in the presence of the typical grey aliens who conduct medical examination. In many cases abductees do not have conscious memories of the event, although subconscious images make their presence felt through dreams. Upon awakening from an induced sleep, subjects report feelings of anxiety. There are often physical manifestations of the experience, such as blood on pillows or bedclothes, bruises that seem to have appeared out of nowhere. Abductees often point to physical evidence that confirms to them that their experiences have been real and not dream-induced.

ALIENS

There is a great deal of skepticism regarding alien abductions, not only within the professional psychiatric community, but within the UFO community as well. Most memories of alien abductions are recovered via regression therapy. While there are traditional abduction reports, there are also bizarre acosmological philosophy, shown the future of planet Earth, and told to go back to report to others what they have seen. Of course, it is very easy to characterize these abductees as delusional because they often have no palpable evidence to support their stories. Despite numerous sightings, suspected crashed saucers, and alleged authentic photographs/ videotapes/ films of true UFOs, ufology has yet to provide anything convincing. Are aliens really here? Are the strange light phenomena natural? Are the unexplained objects flying through the sky alien vehicles? Are abduction accounts real or psychological? Are close encounter stories with flying saucers called by hypnosis? Are luminous phenomena a neuro-sense problem? Are the gray light of balls seen alien encounters? Are these experiences due to hypersensitivity to intense electromagnetic field? These are stunning questions in the minds of people.

Ufology always makes claims, but science does not do research on them. It looks good in the media, but science will always scoff at those theories that have no hard data or facts to support their conclusions. Ufology has spent the last sixty years yearning for some kind of scientific recognition unsuccessfully. However, science has either ignored the subject or rejected the Extra-Terrestrial Hypothesis. There are many people who report seeing or having encounters with aliens. There are five kinds of encounters with aliens: the system of classifying UFO sightings. A close encounter of the first kind is seeing a

UFO in close proximity within 500 feet to the witness. A close encounter of the second kind is a UFO leaving marks on the ground, causing burns or paralysis to humans, frightens animals, or interferes with car engines or TV and radio reception. A close encounter of the third kind is a UFO with visible occupants. A close encounter of the fourth kind is alien abduction cases. A close encounter of the fifth kind is the communication that occurs between a human and an alien being. An extraterrestrial commonly called an alien is a being from another world.

Aliens have become a popular symbol of science fiction over the years. Aliens vary in appearance by many different ways. The most common alien reported by many abductees are the Greys. The Greys are described as short, thin-limbed and slight-bodied, with large heads and huge, almond-shaped black eyes and gray coloured skin. The Greys are also called Zetas since it is believed they come from the twin-star system Zeta Reticuli. They do not communicate via spoken language but instead transmit their thoughts telepathically. Also aliens are said to communicate with Earth people by making shapes in crop fields called Crop Circles. Aliens are usually reported as driving aircraft known as UFOs. It is unkown as to how aliens seat themselves and pilot UFOs. Aliens have also been known to use aquatic vehicles such as USOs which may look like ordinary UFOs. The grey aliens are the ones known for abductions and probing. There is Oz effect associated with UFOs. The physical and neuropsychological phenomenon associated with encounters with UFOs or extraterrestrials in which the intense amount of electromagnetic activity produced by the craft or being slows down the physical and mental processes of the human witness.

The feeling of unreality of a disconnect with his or her physical surroundings, is actually a reaction to the intense electromagnetic or static field generated by the object or UFO. An electromagnetic envelope or cushion of energy that surrounds flying saucers, allows them to mitigate through space, prevents intruders from approaching, and causes people to feel disoriented. Because of the witness's sense of unreality, the phenomenon was dubbed "the Oz effect" by UFO researchers. UFO researchers/scientists conclude that UFOs possibly convert gravitational- field energy for propulsive purposes. It is a method of propulsion that relies on the conversion of gravity, a re-harnessing of a force that exists everywhere in the known universe and is, therefore, inexhaustible and recyclable. Zero Point theory suggests that empty space is really not empty at all, but rather a reservoir of gravity- energy that can be turned on and off by "quantum fluctuations". If this reservoir of endless power could be processed by some sort of engine, it would be theoretically capable of generating enormous speeds that could literally bend the space-time continuum so as to enable travelers to journey through the medium of time as well as space.

These are directed-energy or particle-beam weapons that shoot high-speed streams of electrons at their targets, which can destroy electrical circuitry, disable computer controlled systems, and detonate incoming warheads. Abduction researchers have studied many bizarre experiences and found out certain indicators of possible abductions: missing or lost time, especially an hour or more; unusual scars or marks with no possible explanation; emotional reactions like fear or anxiety or panic; memories of flying through the air; vivid dreams of strange alien creatures or flying objects; experience of strong aversion or attraction towards aliens; blood or unusual stains on bed sheets or pillows; strange humming or pulsing sounds; chronic and untreatable headaches, sinusitis, or nasal problems; frequent or sporadic ringing in the ears; paranormal or psychic experiences; abnormal sensitivity towards certain lights or sounds; compulsive or additive behavior; back or neck problems; trying to resolve these problems with little success; etc. UFO researchers believe that DNA Computer are a part of the genetic makeup of extraterrestrials and control the adaptation of cloned extraterrestrial Being Entities to native environments.

DNA Computer or Molecular Computer replicates DNA and enzymes. The enzyme FOK1 breaks bonds in the DNA double helix, causing the release of enough energy for the system to be self sufficient so that the computing as seeded by an alien race, DNA research holds the promise that their theories might actually be proven correct: aliens have created living species on Earth by manipulating strains of earthly DNA. There are now many hundreds of alleged UFO abduction cases on record, which in ufology are known as Close Encounters of the fourth kind. About a half of them come to light through the use of regression hypnosis, which can also reveal one's inner fantasies just as readily. Not all ufologists embrace it warmly as it not only confuses matter, but also harms witnesses. Some abductees have pieced together their recollections via periods of missing time. The harvesting of genetic material has emerged strongly in the majority of cases from all parts of the world. A detailed study of abduction phenomenon around the world, all share a common base tempered by both individual mythically and culturally derived factors.

British aliens tend to be far more refined, and rather human-like, while American aliens are mostly small, grey-skinned, peer-headed creatures with a cool, clinical style. Alien abductions are undoubtedly a major puzzle for science to resolve, and however, it is unclear where the answers will come from. There is the fantasy/reality incongruity of the UFO subject. Yet it is a proof of something to explore. New ufology proposes a radical theory of an imaginary interpretation of UFO close encounters as an unusual atmospheric phenomena might be capable of interaction with the consciousness of susceptible witnesses and trigger visions of alien visitors. Research psychologists talk of the UFO as a sort of paraphysical creation of the mind, which results from myths and archetypes from the collective unconscious describing other worldly beings. But extensive research shows beyond doubt that the witnesses were sincere and faithfully reported that they believed they had seen. There have been literally thousands of unexplained animal mutilations, mainly in America, Canada, Puerto Rica, Brazil, the Canary Islands, and Spain.

Domestic farm animals have been butchered with surgical precision with their key glands excised and taken away. Most of these have been cattle, although horses, sheep, and dogs have been reported too. These animals have been discovered with internal organs surgically removed, and many of them drained of blood. No blood has been found at the locations, no perpertrator has ever been caught, and no satisfactory official explanation has ever been offered. The animals are often found in the middle of a perfectly created circle. Skeptics and government officials blame natural predators and Satanists. The sheer number of killings defies such obvious explanations as well as no human predator has ever been caught. It is a real mystery, and inevitably one that some believe has a causal link with

UFO phenomenon. There are more revealing cases of alleged communications between human beings and extraterrestrial beings. These fascinating cases show the amazing hidden capabilities of the unconscious mind, which creates psychic fantasies by drawing on whatever it finds suitable for its purpose: this new craze is called channelling.

A channeller is, in truth, little more than a spiritual medium who relays psychic messages from outer space. Their experiences extend into the realms of the UFO phenomena. They become regular outlet and begin to undergo contacts and visionary trips to unknown worlds. They even describe the landscape and the fauna of the other worlds and even learn the names of aliens and creatures of many other planets. The channellers are usually clever and intelligent: their claims may look eccentric or apparently absurd, but there is little doubt about the sincerity of most of them. Many of the channellers are extremely gifted at some form of creativity of visualization, or wonderful imaginations, or exceptional visual acuity, or extrasensory perception. The messages relayed through them, mostly comprises friendly warnings about the mess with our environment and some sort of universal brotherhood of peace. The new phase of UFO research seems to focus not only on the physical and biological aspects of UFO phenomena, but also on the psychological and spiritual aspects of diverse messages or experiences from contactees, abductors, and channellers.

Anyone who starts to obtain information on UFOs and other phenomena may change personality. There are several hundreds of encounter cases where adults revealed, often under hypnosis, memories of being abducted by alien beings. Abductee can be any age and in any environment when they are abducted. They are often rendered immobile by some type of telepathic process and then moved to a waiting alien vehicle which is nearby. After medical and surgical work, the abductee is brought back to the site of their initial abduction. Within seconds of being returned to their abduction site, the abductee awakens and remembers nothing that has just happened to him. The abduction report cases possess clear signs of manipulation of space and time and witness consciousness. The reports of abduction cases are all very similar and contain details of the appearance, actions, and interactions of the small humanoids, their space vehicles and even the surgical utensils used in the physical examinations that are the central theme of these memories. The striking elements of the abduction memories are the initiation of transforming, consciousness expanding phenomenon, and the eventual development of relationships with alien beings.

A large number of abduction experiences recalled through hypnosis give credence to a larger hybrid or interbreeding program. People who encountered aliens have stated that these beings communicate through telepathy. There is no need for words when they can communicate mind to mind. We do not know that those interacting with humanity via abductions have a negative orientation and intend to create race of human-alien hybrids capable of ruling over humanity. The nefarious extraterrestrial agenda to systematically exploit and enslave humanity seems as an attempt to gain control of the planet and its resources. Some happen to involve highly risky and dangerous genetic experiments for a variety of nefarious purposes. Abductees refer to seeing hybrid toddlers, teens, and even adults who closely resemble humans that they are apparently living among us. Alien abductions are undoubtedly a major puzzle for science to resolve and we are unclear about answers. However, trying to scientifically prove the existence of extraterrestrial beings does not require an intimate understanding of their origin, purpose, and significance.

We do not know the world would be a better or less stable place after the great event of alien contact. The consequences of such a discovery of intelligent life are presently unpredictable. But the chances of mass panic seem unlikely as millions of people already believe in extraterrestrials. Perhaps it would be difficult to cope with having to accept into our society, a whole new basically humanoid, but odd-looking extraterrestrials. Some even fear that the sudden confirmation of superior alien intelligence would lead to a traumatizing effect on our society. One is bemused with the idea that alien contact would be the destruction of political demagogues as they bring empires crashing down. It may be that alien contact would produce financial chaos, shattered religious beliefs and cultural allegiances. Others suggest that aliens might want to prepare for our future by gradually adding knowledge or influencing our beliefs. There are people who believe that we do not realize consciously that we have already had alien conditioning, or even have alien heritage ourselves. In the light of growing social acceptance that we are not alone, aliens would just be accepted.

But most ufologists reach the inevitable conclusion, when the extraterrestrial community arrives, there would be the homely feel of this happy band and invite them as Ufolks. One would hope that there is a real space brotherhood possible and they might expect us to try to help ourselves. The inability of science to rid the world of evils, may have prompted the idea that we need to look beyond the mundane for rescue. Probably, the alien agenda seems to create a hybrid species, designed to appear more human like and successfully manipulate science and culture in preparation for their eventual takeover of the planet. Ufology is not embraced by academia as a scientific field of study, and is instead considered a pseudoscience by skeptics. Pseudoscience is a term that classifies arguments that are claimed to exemplify the methods and principles of science, but do not in fact adhere to an appropriate scientific method. It lacks supporting evidence, plausibility, falsifiability, or otherwise lack of scientific status. Scientific UFO research is challenged by the facts that the phenomena are spatially and temporarily unpredictable, are not reproducible, and lack tangible physicality.

CHAPTER TEN

Alien Science

"Unidentified aerial phenomena, otherwise known as
UFOs are real, not the stuff of science fiction
- Leslie Kean

The presence of UFOs in the universe has been ignored long enough: even Columbus had a UFO sighting. Throughout the last sixty tears period, UFOS have been actively observed in our civilization. In fact, UFO sightings have now become so common, yet we have no time to worry about. Whenever a UFO appears, we all simply ignore it. There are innumerable exhaustive reports from people claiming that they had contact with a spacecraft full of aliens. UFO sightings are not limited to farmers in rural areas, but many astronomers, astronauts, pilots, and military personnel have witnessed events. UFOs are picked up by ground and air radars, and they have been photographed by gun cameras all along. There are so many UFOs in the sky that the Air Force need to employ special radar networks to screen them out. The facts about UFOs have long been tracked down and results have long been known in top secret defense circles of many countries around the world. The matter has been the most highly classified subject in the superpower countries. There is another aspect to the UFO phenomenon that involves politics and secrecy rather than observational evidence.

Governments seem to usually ignore UFOs, and routinely issues false explanations. It seems hardly possible that all these reports could be due to optical illusions. Probably there exists government sconsideration. Also intelligence agencies, security agencies, and public agencies are involved in the cover up of facts pertaining to the situation. But the government's pronouncements based on unsound statistics serve merely to misrepresent the true character of the UFO phenomena. There exists a vast amount of high quality, albeit, enigmatic data. But any possible evidence of such in the form of a subset of UFO is ignored or ridiculed. Some UFOs are extraterrestrial vehicles and some of them have crews, while some of those crews catch and release humans to study. Besides we have so much data such as Crop Circles and cattle mutations, alien abductions, and human-alien hybrids. Most people around the world believe that they have been visited by aliens. To back their claims, witnesses have presented snapshots of flying saucers and debris from crash landing.

The over-all public and scientific response to the UFO phenomenon is itself a matter of great scientific interest, above all in its social-psychological aspects. However, most scientists reject outright the validity of UFO research, refuse to engage in it, and deliberately ignore the intriguing data compiled by UFO fraternity. For many years scientists have avoided discussion about their hypothesis and research surrounding the top of UFO investigations. Most of them have absolutely no idea where the UFOs come from or how they are operated, but they know that they are something to come from outside our atmosphere. Many scientists have been reluctant to come forward up until now with regard to UFOs due to fear of backlash. Most

UFO skeptics are quick to dismiss as impossible idea that UFOs are alien spacecraft. But many top scientists now believe and accept the probability that UFOs do exist. The questions about extraterrestrial life have left the realm of theology and science fiction, and entered the realm of experimental science. Our best modern astrophysics theories predict we should be experiencing extraterrestrial visitations.

In September 2013, a team of British scientists claimed a cell fragment found by a balloon flight in the upper atmosphere may be proof of life form in space. We become obsessed with the notion of alien life forms or strange array of aliens. UFO sightings represent a solid, physical phenomenon that appears to be under intelligent control, and is capable of very high speeds, maneuverability, and luminosity beyond current known technology. Scientists often state that some of the phenomena related to UFOs are not easy to explain in the conventional fashion. There is so much to explore in this fine study of UFOs. Some UFO researchers make their point clear that UFO might have an impact on every person on the Earth and therefore we should learn about it. There lacks a plausible conventional explanation to the UFO phenomenon as it is both complex and dangerous. The UFO problem is not a simple one, and it is unlikely that there is any simple, universal answer. The subject of UFO is one that is not yet known and not usual so that scientist can get back in using scientific method to study things. It is beyond doubt that the UFOs have an extraterrestrial origin.

They are surely artificial objects under intelligent control. They are definitely the craft of a supremely advanced technology. These objects are interplanetary craft of some sort. UFOs are undoubtedly conceived and directed by intelligent beings of very high order. The UFO problem is a multi-situational and multi-dimensional phenomenon. Alien crafts are from both ultra- dimensional sources and sources within our dimension. So much data have been gathered which has often pointed to aspects of the phenomena that have been suppressed. There have been so many reports from responsible observers that UFOs cannot be ignored. Of particular concerns are reports that UFO encounters may be hazardous to people's health. Some sightings that are accompanied by physical evidence deserve scientific study. Science needs to focus on incidents involving some form of physical evidence including photographic evidence, radar evidence, vehicle interference, interference with aircraft equipment, apparent gravitational or inertia effects, ground traces, injuries to vegetations, physiological effects on witnesses, human radiation-type injuries, and debris.

The language transmutation is common in UFO narratives such as hovering motionless, making no noise, accelerating to very high speeds, transforming into speed off in the blink of an eye- such statements making it harder for scientists to provide natural explanation. Although UFO reports date back 60 years, the information gathered does not prove that either unknown physical processes or alien technologies are implicated. Nevertheless, they do include a sufficient number of intriguing and inexplicable observations. Most current UFO investigations lack the level of rigour required by the scientific community, despite the initiative and dedication of the investigation authority. More work needs to be done to bring all UFO anomalies into the accepted paradigm or adjusting our paradigms based on evidence. Scientific evaluations must take place with a spirit of objectivity and a willingness to evaluate rival hypotheses. Scientifically acquired data and analysis could yield useful information and advance our understanding of the UFO problems. But unexplained observations, brings the possibility that scientists will learn something new surely by studying them.

Ufology is the study of unidentified flying objects and there is a need to bridge the gap between science and ufology. Otherwise, we would miss those strange and possibly breakthrough attributes of the UFO phenomenon, if our insights into the UFO phenomenon was the debunking official military response revealed time and time again by their documents and histories. The hypothesis that UFOs are of extraterrestrial or inter-dimensional origin is definitely a rational one. There are numerous theories on how UFOs travel through the deep dark regions of space and the atmosphere of Earth. Experts suggest the use of anti-matter technology, ion propulsion, gravity drives, or they have even developed a method to open up wormholes through space and time. Alien technology seems to include advanced propulsion systems, energy systems, antigravity, new metals and alloys, time travel, light bending techniques, weaponry based on beams, computing techniques, and teleporting techniques. Alien travelers arrive here on Earth from distant star systems via a flying disc shaped object. To achieve this either alien beings life span is likely to be extremely long or their advanced knowledge has allowed them to become immortal.

UFO propulsion is comprised of an advanced electromagnetic field, which would explain why people experience their cars as suddenly stopping or their radios going all screwy. Some UFO researchers suggest that UFO engineers had managed to develop a 'field engine'. The essence of the field engine is a static field link between the UFO and the earth or other large mass of planetary or stellar. Aliens possibly have engineered their UFO spacecraft to solve the problems associated with travelling at the speed of light or even faster. As an object approaches the speed of light, time begins to slow down. The principle of time dilation states that bodies moving at high speed of light, experience a time that ticks slower than the time measured at zero velocity. Alien technology reports reveal that alien spaceships contain controls that interface directly with the pilot's consciousness. In other words, it means that the craft is designed especially for the consciousness of the alien pilot and may not necessarily be suitable for another human to use. It also means that an alien entity and its interstellar vehicles are one. It looks alien technology is spiritual in nature.

Space-time consists of points or events that represent a particular place at a particular time. Wormholes are holes in the fabric of four-dimensional space-time, which are connected, but originate at different points in space and at different times. They provide a quick path between two different locations in space and time. It has been reported that the people inside a UFO do not sense any movement when the craft moves from one location to another. No matter how fast the craft moves or how many times it changes direction, the people inside feel as though the craft is completely stationary. Aliens are also capable of manipulating light, where a beam of light shown towards a UFO is bent at a ninety degree angle away. Humanity is not yet fully ready to accept the existence of aliens and UFOs. The alleged aliens are believed to be really not aliens as in coming from other planets or stars but rather inter-dimensional beings which come from a parallel universe or reality. For those people who have the consciousness for higher reality, the evidence of aliens will manifest with greater effect. Those that have the consciousness to perceive the reality of aliens can interpret the evidence as real.

Our opposing collective consciousness creates a half way effect where the manifestation of evidence can be true or false. So the beliefs play a major part since the existence of aliens and UFOs are very much mental in nature. The secret of faster than light travel is that it takes place in the nonphysical dimension of the universe. Alien spacecrafts are usually invisible as they observe human activities, because they view physical reality while hovering in the etheric plane. When they move into our plane momentarily, they can seem to appear out of nowhere. When they move back into the etheric plane, they seem to disappear without a trace. UFOs behave sensibly as in group-formation flight, they maintain a pattern. They are most often spotted over engineering installations, atomic stations, and airfields. They always maneuver so as to avoid direct contact, while encountering aircraft. Such intelligent actions of UFOs give the impression that they are investigating or perhaps reconnoitering. Lasers and directed particle-beam weapons incorporated into some highly regarded anti-ballistic missile defense systems have been derived from the reverse-engineering of UFO craft.

The value of laser comes from its almost limitless range and high power, and its ability to acquire targets, point, and shoot, and reacquire in mere seconds. Stealth Bomber: the crescent-shaped, radar-evading, flying-wing-type aircraft reportedly have a shape similar to UFO that crashed in Rosewell. The most top-secret military facility in America is Area 51 - in Nevada: the base is so important to American military aerospace programs. It is said to contain hangars and laboratories to store and study alien saucers. It has supposedly stored and maintained not only flying saucers but actual extraterrestrials. The Science of Aliens- a touring exhibition was launched at the London Science Museum in October 2005. With a combination of artifacts, interactive and audiovisual exhibits, the exhibition dealt with the real possibilities for alien life, explored the variety of life on Earth, displayed alien worlds documentaries, and concluded by looking at the chances of communication with alien intelligences. Wormholes and hyperspace are not well-known concepts in the scientific community. Hyperspace consists of those dimensions which are co-dimensional with our space-time.

Real alien teleportation means that they could be transported to any location instantly, without actually crossing a physical distance. Teleportation is the theoretical transfer of matter from one point to another without traversing the physical space between them. Teleportation involves dematerializing an object at one point, and sending the details of that object's precise atomic configuration to another location, where it will be reconstructed. This means that time and space could be eliminated from travel. It was then in March 1993, that physicist Charles Bennett and a team of researchers at IBM confirmed quantum teleportation was possible, but only if the original object teleported was destroyed. Astro-theologists feel that the study of the astronomical origins of religion does hope to tie those odd UFO encounters and sightings to something far more complex than advanced civilizations in far corners of the cosmos. Some of the world's leading astronomers believe that aliens may be using an entirely different communication medium such as ghostly neutrinos or gravitational waves which are cripples in the fabric of space-time, or using communication mechanisms we cannot begin to fathom.

Our 21st- century technology seems too primitive to detect advanced extraterrestrial life. We need to ponder over the implications of making contact with extraterrestrials and its impact on public policy and the future of our planetary life. It is apparent that there is a very real and quite large gap existing between the alien science and the science in which we have been trained. Certain crucial experiments that have been suggested and carried out, showed to confirm the validity of the alien science. Beyond certain point the alien science just seems to be incomprehensible. The US government has had live alien hostages at some point in time and has conducted autopsies on alien cadavers. Early US government efforts at acquiring alien technology were successful. The secret government effort has super technology that mostly came from aliens or from back engineering downed alien spacecraft. They are suspected to have working relationship with alien forces for some time, with the express intent of gaining technology in gravitational propulsion, beam weaponry, and mind control. Anti-gravity is the idea of creating a place or object that is free from the force of gravity.

It does not refer to countering the gravitational force by opposing force of a different nature. Instead, anti-gravity requires that the fundamental causes of the force of gravity be made either not present or not applicable to the place or object through some kind of technological intervention. The basis of our genetic development and religions lies in the intervention by non-terrestrial and terrestrial forces. Some scientists believe that we have alien DNA in our genetic code and that human DNA was encoded with an extraterrestrial signal by an ancient alien civilization. They also claim that human DNA is ordered so precisely that it reveals an ensemble of arithmetical and ideographical patterns of symbolic language. We can never take the credit for our record advancement in certain scientific areas alone as we have been helped by aliens. The famous Albert Einstein and Nicola Tesla were both said to have alien connections. It is always productive to continue contact between the UFO community and our physical scientists that helps find ourselves part of a large, galaxy-sized civilization. Perhaps, we should consider the possibility of being helped by advanced extraterrestrials from outer space.

A lot of scientific theories we have developed failed to take into account the higher dimensional characteristics and properties of the subjects. We need to improve our observational skill of the universe, understand nature, adopt natural forces to our living benefits, and make energy work for us. Leading cosmologists believe the existence of extraterrestrial life may be beyond human understanding. There could be life and intelligence out there in forms that we cannot conceive or aspects of reality that are beyond the capacity of our brains. If other life forms are out there and visitors have come, then they might have brought their alien technology with them. The challenging task of knowing aliens from other worlds is useful for any species that dreams of understanding its place in the universe. There is a scientifically-based speculation at the far edge of knowledge-and beyond: the day will arrive soon when we people on the Earth will finally admit we are not alone in the universe, and we humans will come to face with other life forms from the cosmos. There is likely a current active presence on our planet among us that controls different elements of our society.

Also facts indicate alien overt presence within a decade. We do not know the nature of extraterrestrial civilization as no conclusive contact with any has yet taken place. Therefore, it is impossible to say with complete accuracy what the result of contact would be. The medium through which humanity is contacted, be it electromagnetic radiation, direct physical interaction or perhaps an extraterrestrial artifact, could also influence the result of contact. The impact of alien invasive species are immense, insidious, and usually irreversible. Also the results of extraterrestrial contact depend on the method of discovery, the nature of their beings, and their location relative to the Earth. The resulting changes from extraterrestrial contact would vary greatly in magnitude and type based on their civilization's benevolence or malevolence, its level of mutual comprehension between itself and humanity. One type of extraterrestrial, the Annunaki, left a huge legacy on human biology and culture. Most likely it could give human access to a galactic heritage, perhaps predating the human race itself, which may greatly advance our technology and science.

Alien civilization which is far ahead in technology than humanity, have been likened to the meeting of two vastly different human cultures on Earth. The discovery of extraterrestrial intelligence will be crucial for the humanity as it will affect life on Earth and change the philosophical consciousness of human. The alien contact could have a profound impact on religious doctrines, potentially causing theologians to reinterpret scriptures to accommodate the new discoveries. It might make us see ourselves as more of one planet and less a collection of separate societies. The world, quite simply, would never be the same again. During the last half of twentieth century, rapid advancements in science and technology prompted many people to begin rethinking our place in the universe. If aliens reach out to us, it is a question that has puzzled science-fiction fans and scientists alike for decades. Would they be friendly explorers, or destroyers of worlds? The potential changes from extraterrestrial contact could vary greatly in magnitude and type, based on their level of technological advancement, degree of enevolence or malevolence, and level of mutual comprehension between itself and humanity.

CHAPTER ELEVEN

Alien Enigma

"Sometime I wish the aliens would abduct me and crown me as their leader"- George Noory

The existence and visitation of highly intelligent beings is probably one of the most captivating mysteries of our time. If intelligent Aliens visit the Earth, it would be one of the most profound events in human history. Unidentified Flying Objects are more often than not associated with Aliens, since legends have it that these flying objects are being flown by again someone or something. There are myths and legends, stories and folklores, all woven around the experiences of sighting something in the sky. The number of UFO sightings is currently flying at an all-time high, and there have been more than 1,00,000 recorded UFO sightings in the past 100-plus years. Nearly half of Earthlings believe that we are not alone in the universe, and feel sure that Aliens have visited Earth, either in the ancient past or recently. Many of us are fascinated with the idea of extraterrestrial intelligence and UFOs have captured our imagination. Scientists do not deny the existence of intelligent Aliens, but dismiss these beings as not representing real physical phenomena. They do not discount the idea of Aliens, but they are not convinced by the evidence to date, because it is undeniable.

ALIENS

If scientists are not convinced by the current evidence of UFOs, which does not mean that they do not exist. Scientists have been monitoring radio waves for signs of Alien life in the universe for decades, and they have not found anything or anyone. The absence of evidence for intelligent Aliens is called Fermi Paradox. In the light of recent Exoplanet discoveries, it makes very unlikely that we are the only, or the first advanced civilization. Astronomers use the Drake Equation to estimate [i]number of technological Alien civilizations. The Search for Extra-Terrestrial Intelligence or SETI and Messaging Extra-Terrestrial Intelligence or METI have been looking for signs of intelligent life elsewhere In the universe for a long time, but the search has so far found nothing. Searching for extraterrestrial life is no easy feat. Scientists and astronomers express varying degrees of enthusiasm for the possibility of intelligent life in the universe. Alien visitation is more strongly correlated with the paranormal belief measures rather than the extraterrestrial belief factors. It may well be that while the twentieth century was the century of physics, the twenty-first century will be about astrobiology.

The big question, however, is what that life might look like, or what Aliens might be like. Scientists believe that laws of the universe are the same everywhere, then different descriptions of these laws should in principle, be equivalent. Some astrobiologists understand that Aliens are potentially shaped by the same processes and mechanisms that shaped human, such as natural selection. The theory supports the argument that foreign life forms undergo natural selection, and are like us, evolving to be fitter and stronger over time. Like humans, they predict that they are made-up of a hierarchy of entities, which all cooperate to produce Alien. When it comes to language, which is the single most important factor in human cooperation, we do not know whether it is plausible that any technologically versatile Alien civilization would have something like language. However, a direct meaningful communication with extraterrestrial intelligence is highly improbable. If Aliens ever do touch down to Earth, their language is likely to challenges not found in any human language. Even our own biology could limit us from understanding an Alien language.

The discovery of Intelligent Aliens could make believers feel insignificant, and as a consequence cause people to question their faith. Contact with intelligent life elsewhere in the universe will present theological and philosophical conundrums. The core question could be does God's creation extend beyond a single planet. If so, would the inhabitants of those planets believe in the same Gods as humans do. Exotheology is a term that describes theological issues related to extraterrestrial intelligence. There is no need to imagine that God reveals the same truths in the same way to all intelligent life in the universe. Other civilizations could understand the Divine in their own myriad ways. Alien races may be highly advanced that humans would be seen as barely intelligent apes if compared with them. Astrophysicists suggest that advanced and ancient civilizations may exist but be beyond our comprehension or ability to detect. We do not really know, we were the first technological species on Earth, if some earlier civilizations existed on Earth millions of years ago, we might have trouble finding evidence of it.

Ever since UFOs were first spotted in the sky in 1947, many conspiracy theories have sought to explain these apparitions that defy science. A whole host of conspiracy theories have made their way into the mainstream over the last few decades, but we may have to learn the full truth, amid all the cacophony. There is a specific day dedicated for UFOS: July 2. UFO technology is based on some remarkable and rather exotic physics. Reports of Unidentified Aerial Phenomena should be the subject of serious study and scientists should at least take the notion seriously. It is theoretically possible that this has happened, not entirely impossible, and work looking for evidence that it has. Advanced Aliens have transcended their physical form with tremendous computational power available with super-intelligence quicker than humans. There is a good chance that extraterrestrial life will look like us, but with cognitive sophistication, large brains, and intelligence. Whether the answer to the question is there anybody out there? is yes or no learning the truth will completely change humanity's understanding of our place in the universe.

There are so many forms of energy in the universe and Aliens are not magic. Researchers believe that although life appears in many forms, the scientific principles remain the same. Scientists also observe Alien evolution may well produce form of intelligence that as far superior to ours and no longer based on carbon. Advanced Aliens could have evolved their physical form to become super intelligent AI beings. They propose that Alien life could be advanced that it becomes indistinguishable from physics. Some science experts believe Aliens do exist among us, but in a form that humans cannot see it. The shadow biosphere theory postulates that other forms of life may exist alongside known organisms, but operate in ways we do not recognize them. If Alien minds are entirely different from ours, communication might be impossible. They may, for instance, be able to tap into the energy released by super-massive black holes or active galaxies known as quasars. Some feel If Aliens ever visit us, the outcome would not turnout very well to Earthlings, as they are very powerful, and could be aggressive. Others say that their visit to Earth would benefit us to know them, their nature, and their intent.

Bibliograpbhy

A.Thom: " Megalithic Lunar Observations"- Oxford University Press 1971.

Aime Michel: "The Humanoids"-Henry Regney 1969.

Alan Watts: "UFO Visitation"- Bland ford Press.

Allen Hendry: "The UFO Handbook"- Sphere 1980.

Allen Hynek: "The UFO Experience"- Abelard-Schuman 1972.

Amanda Davis: "Extraterrestrials: Life in Outer Space"- Rosen Pub Group 1997.

Anna jamerson: "Connections: Unraveling our Alien Abduction"- Wild Flower Press 1996.

Anthony Lawtson & Jack Stonely: "Is Anyone Out There"- W. H. Allen 1975.

Beth Davis: "Ciphers in the Crops"- Gateway Books 1992.

Bill Barry: "The Ultimate Encounter"- Pocket Books 1978.

Bob Larson: "UFOs: And the Alien Agenda"- Thomas Nelson Publishers 1997.

Bob Rickard: "UFOs"- Gloucester Press 1979.

Bob Shaw: "The Ragged Astronauts"-Gollancz 1987.

Brad & Francis Steiger: "The Star People"- Berkley 1981.

Brad Steiger: "Alien Meetings"- Ace 1978.

Brian O Leary: "Exploring Inner and Outer Space"- North Atlantic Press 1989.

Carl Jung: "Flying Saucers"- Routledge & Kegan Paul 1959.

Carl Sagan: "Communication with Extraterrestrial Intelligence"- MIT Press 1973.

BIBLIOGRAPBHY

Carl Sagan: "contact: UFOs- A Scientific Debate"-Norton 1972.

Charles Berlitz & William Moore: "The Roswell Incident"-Grafton 1980.

Charles Cazean: "Exploring the Unknown"- Plenum Press 1979.

Clifford Wilson: "The Alien Agenda"- Signet Books 1988.

Clifford Wilson: "UFOs and Their Mission Impossible"- Signet 1974.

Daniel Cohen: "A Close Look at Close Encounters"- Dood Mead 1981.

David Holmes: "The Search for Life on Other Worlds"- Bantam 1966.

David Langford & John Grant: "Earthdom"- Grafton 1987.

David Seargent: "UFOs: A Scientific Enigma?"- Sphere 1978.

Dennis Bardens: "Mysterious Worlds"- Cowles Book 1970.

Diane Tessman: "The Transformation"- Inner Light 1988.

Donald Keyhole: "Flying Saucers: Top Secret"- G.P. Putnam's Sons 1960.

Donna Leatherberry: "UFOs Exposed in Scriptures"- Northwest Pub Inc 1995.

Edward Ashpole: "The Search for Extraterrestrial Intelligence"- Blandford 1989.

Eric von Daniken: "Chariots of the Gods?"- Souvenir 1969.

Erich von Daniken: "Gold of the Gods"- Corgi 1975.

Erich von Daniken: "In Search of Ancient Gods"-Corgi 1976.

BIBLIOGRAPBHY

Erich von Daniken: "Miracles of the Gods"-Pine Brook Dell 1975.

F.Petric: "Wisdom of the Egyptians"- Quaritch 1940.

Fred Hoyle: "Life Cloud"- Dent 1978.

George Adamski: "Behind the Flying Saucer Mystery"- Warner Paperback 1974.

George Andrews: "Extraterrestrials Among us"- Llewellyn Pub 1992.

George W. Cox: "Alien Species and Evolution"- Island Press 2004.

Gorden Allen: "Spacecraft from Beyond Three Dimensions"-Exposition Press 1959.

Greg Bear: "The Forge of God"- Gollancz 1986.

H.P.Lowercraft: "The Haunter of the Dark"- Grafton 1988.

Hans Holzer: "The UFONAUTS"- Fawcett 1976.

Ian Ridpath: "Messages from the Stars"- Fontana 1978.

Ian Watson: "Miracle Visitors"- Gollancz 1978.

Isaac Asimov: "Fantastic Voyage"- Bantam Book 1982.

Isaac Asimov: "Unidentified Flying Object"- Gareth Stevens 1988.

Jacques Valley: "UFOS in Space; Anatomy of a Phenomenon"- Random house 1965.

James Mosley:"UFO Crash Secrets"-Abelard Productions 1991.

James Oberg: "UFOs and Outer Space Mysteries" – Donning 1982.

Jean Sendy: "The Coming of the Gods"- Berkeley 1973.

Jeffrey Bennett: "Beyond UFOS"- Princeton University Press 2008.

Jenny Randles & Paul Whetnall: "Alien Contact"- Coronet 1983.

Jenny Randles: "The UFO Conspiracy"- Blandford 1987.

Jerome Clark & Loren Coleman: "The Unidentified"- warner 1975.

Jim Lovenzen: "Abducted"- Berkeley Books 1977.

John G. Fuller: "The Interrupted Journey"-Souvenir 1980.

John Grant: "Dreamers"- Ashgrove 1984.

John Grant: "Great Mysteries"- Chartwell 1989.

John Keel; "UFOS: Operation Trojan Horse"-Putnam 1970.

John Magor: "Our UFO Visitors"- Hancock House !977.

John Michel: "The Evidence for Alien Abductions"- The Aquarian Press 1984.

John Spencer: "Perspectives"-Macdonald 1990.

Josef Blumrich: "The Spaceships of Ezekiel"- Bantam 1974.

Kal K. korff: "Spaceships of the Pleiades"- Prometheus 1996.

Larry Kettlekamp: "ETs and UFOS: Are They Real"- William Morrow & Co 1996.

Leonard Cramp: "Space, Gravity and the Flying Saucer"- Werner Laurie 1954.

Mark Eastman: "Alien Encounters"- Harvest House Publishers 1997.

Maurice Chatelan: "Our Ancestors Came From Outer Space"- Pan 1980.

Max Toth & Greg Nielsen: "Pyramid Power"- Warner 1976.

Neil Tonga: "Alien Encounters"- Sterling Publishing 1998.

Neugebauer: "The Exact Sciences f Antiquity"- Princeton 1951.

Nigel Blundell & Roger Boar: "The World's Greatest UFO Mysteries"- Octopus 1983.

Paris Flammonde: "UFOS Exist"- Putnam 1976.

Patrick Tilley: "Fade Out"-Sphere 1989.

Paul Davies: "God and the New Physics"- Penguin 1983.

Paul Davies: "Other Worlds"- Dent 1980.

Paul Davies: "The Cosmic Blueprint"- Heinemann 1989.

Peter Hough & Jenny Randles: "Looking for the Aliens" -Brandford Book 1991.

Peter Kolosimo: "Spaceships in Prehistory"- University Books 1976.

Philip Klass: "UFO Abductions"-Prometheus 1989.

Raymond Drake: "Gods and Spacemen of the Ancient Past"– Signet 1974.

Richard Cavendish: "Man, Myth, and Magic"-Purnell 1970.

Robert Chapman: "UFO- Flying Saucers over Britain" Granada 1969.

Robert Emenegger: "UFOs Past, Present and Future"- Ballantine 1974.

Robert K.G. Temple: "The Sirius Mystery"- Futura 1976.

Ruth Montgomery: "Aliens Among Us"- Fawcett Crest 1985.

Steven Spielberg: "Close Encounters of the Third Kind"- Dell 1977

Stuart Holroyd: "Alien Intelligence"- David & Charles 1979.

Terence Meaden: "The Circle Effects and its Mysteries"- Artetech 1989.

Thomson, J. Eric: "The Rise and Fall of Maya Civilization"-Norman 1954.

BIBLIOGRAPBHY

Timothy Good: "Above Top Secret"- Sidgwick & Jackson 1987.

Tompkins, Peter: "Secrets of the Great Pyramids" Harper& Row 1971.

Walter Sullivan: "We Are No Alone"- Hodder & Stoughton 1965.

Williamson G.H: "Secret Places of the Lion"- Destiny 1983.

Wliiam J. Birnes: "UFO Encycopedia"- The UFO Magazine 2004.

Author Bio

Prof. RVM. Chokkalingam is a former lecturer/curator/scientist and now @ 77 is a local professor living in Bangalore, and engaged in the study of Aliens and UFO phenomenon. He is a Science Museum Scholar, specialized in 'Design of Science Exhibit' @ Science Museum, London. He is a science communicator of repute and made lifetime contribution in the Public Engagement with Science for over 50 years. He has established Museum @ National Aerospace Laboratories, Bangalore in 1999. He is a popular science writer and published more than 150 articles in newspapers and magazines, and authored more than a dozen books. He is the recipient of Karnataka State Award for 'Science Communication' in 2012.

www.ingramcontent.com/pod-product-compliance
Lightning Source LLC
Chambersburg PA
CBHW030838180526
45163CB00004B/1370